Population, Food, and Rural Development

international union
for the scientific study
of population

The International Union for the Scientific Study of Population Problems was set up in 1928, with Dr Raymond Pearl as President. At that time the Union's main purpose was to promote international scientific co-operation to study the various aspects of population problems, through national committees and through its members themselves. In 1947 the International Union for the Scientific Study of Population (IUSSP) was reconstituted into its present form. It expanded its activities to:

- stimulate research on population
- develop interest in demographic matters among governments, national and international organizations, scientific bodies, and the general public
- foster relations between people involved in population studies
- disseminate scientific knowledge on population.

The principal ways through which the IUSSP currently achieves its aims are:

- organization of worldwide or regional conferences operations of Scientific Committees under the responsibility of the Council
- organization of training courses
- publication of conference proceedings and committee reports.

Demography can be defined by its field of study and its analytical methods. Accordingly, it can be regarded as the scientific study of human populations primarily with respect to their size, their structure, and their development. For reasons which are related to the history of the discipline, the demographic method is essentially inductive: progress in the knowledge results from the improvement of observation, the sophistication of measurement methods, the search for regularities and stable factors leading to the formulation of explanatory models. In conclusion, the three objectives of demographic analysis are to describe, measure, and analyse.

International Studies in Demography is the outcome of an agreement concluded by the IUSSP and the Oxford University Press. This joint series is expected to reflect the broad range of the Union's activities and, in the first instance, will be based on the seminars organized by the Union. The Editorial Board of the series is comprised of:

Ansley Coale, USA   Henri Leridon, France
John Hobcraft, UK   Richard Smith, UK
Georges Tapinos, France

Population, Food and Rural Development

Editors
RONALD D. LEE
W. BRIAN ARTHUR
ALLEN C. KELLEY
GERRY RODGERS
T. N. SRINIVASAN

CLARENDON PRESS · OXFORD

Oxford University Press, Walton Street, Oxford OX2 6DP

Oxford New York Toronto
Delhi Bombay Calcutta Madras Karachi
Petaling Jaya Singapore Hong Kong Tokyo
Nairobi Dar es Salaam Cape Town
Melbourne Auckland
and associated companies in
Berlin Ibadan

Oxford is a trade mark of Oxford University Press

Published in the United States
by Oxford University Press, New York

© IUSSP 1988

First published 1988
First issued in Clarendon Paperbacks 1991

British Library Cataloguing in Publication Data

Population, food and rural development.—
(International studies in demography).
1. Developing countries. Food. Production.
Effects of population growth. Effects on food production
I. Lee, Ronald D.
338.1'9'1724
ISBN 0–19–828391–1

Library of Congress Cataloging in Publication Data
Population, food, and rural development/editors, Ronald D. Lee . . .
[et al.].
p.   cm.—(International studies in demography)
1. Population.   2. Food supply.   3. Rural development.   I. Lee.,
Ronald Demos, 1941–   .   II. Series.
HB885.P64 1988   304.6—dc19   88–19627
ISBN 0–19–828391–1

Printed and bound in Great Britain by
Biddles Limited, Guildford and King's Lynn

Contents

Contributors

W. BRIAN ARTHUR   Food Research Institute, Stanford University

PRANAB BARDHAN   Department of Economics, University of California, Berkeley

HANS P. BINSWANGER   Employment and Rural Development Division, The World Bank, Washington DC

MEAD CAIN   Centre for Policy Studies, The Population Council, New York

GRAHAM CHIPANDE   Economics Department, Chancellor College, University of Malawi, Zomba

ROBERT E. EVENSON   Economic Growth Centre, Yale University

ALLEN C. KELLEY   Department of Economics, Duke University

AZIZUR R. KHAN   Country Policy Department, The World Bank, Washington DC

RONALD D. LEE   Graduate Group in Demography, University of California, Berkeley

HANS LINNEMANN   Free University, Amsterdam

GEORGE MARTINE   International Labour Office, Brasilia

JOHN McINTIRE   The World Bank, Washington DC

GEOFFREY McNICOLL   Centre for Policy Studies, The Population Council, New York

PRABHU PINGALI   International Rice Research Institute, The Philippines

GERRY RODGERS   International Institute for Labour Studies, Geneva

MARK R. ROSENZWEIG   Department of Economics, University of Minnesota

T.N. SRINIVASAN   Department of Economics, Yale University

Introduction

Introduction

Population, Food, and Rural Development

RONALD D. LEE

Graduate Group in Demography, University of California, Berkeley

In Malthus's theory of population, the purported inability of a growing labour force to maintain sufficient production of food per head played a central role; returns to labour in agriculture were held to diminish sufficiently rapidly to limit sharply the potential population size under any given state of technology and endowment of natural resources. Likewise today much of the concern about the size and rate of growth of the population, particularly the population of Third World countries, reflects fear that the pressure on land will render it impossible for the agricultural sectors to produce sufficient food. At the same time, others have argued that population growth and higher density induce positive responses which may more than offset the decrease in land per head. This volume addresses these issues and concerns. It begins with an assessment of forecasts of food adequacy at the start of the twenty-first century, with particular attention to their methodological underpinnings. It then considers in more detail some of the processes which are explicitly or implicitly included in the forecasts and many others which are not. For example, how do population density and growth affect agricultural technology, market institutions, and investment? How does population pressure affect access to the land and the growth of a rural proletariat? To what extent is the real problem one of access to the complementary inputs—capital and technology—needed to make the land productive? To what extent do emigration and an open frontier serve to alleviate the problems of rapid growth? These issues are discussed both in general terms and in the context of specific developing countries in Africa, Latin America, and Asia.

The volume begins with T. N. Srinivasan's assessment of current views and projections related to this issue and a comment on this assessment by Hans Linnemann. Srinivasan notes that detailed estimates by the Food and Agriculture Organization (FAO) of carrying capacity for microregions of less-developed countries (LDCs) have limited value, because they take no account of the economic processes which lead to a particular output: they ignore the question of what incentives would be necessary to induce changes in agricultural production, and ignore possibilities of international trade and specialization in agricultural commodities. Those estimates, therefore provide no insight into the trajectories of agricultural investment, technology and output that are to be expected. Other attempts to forecast food output are more comprehensive, however; these range from simple trend extrapolation to structural models. All forecasts agree that food output per head on a global level should be more than adequate through the end of the twentieth century, allowing moderate growth in the quantity and quality of consumption. The most sophisticated forecast searches iteratively for a world price of grain, to which each country's output responds. In this model, experiments with slower and faster population growth were tried; as in the other models, it was found that slower population growth would lead to higher per capita food supplies. One scenario, in which non-agricultural income was assumed to grow slowly, actually led to declining per capita incomes and therefore to declining food demand and food production per head.

To say that food output will probably allow rising per capita consumption over the next two decades is not to say that everyone will have sufficient food. As in the present, there may be important regional imbalances, which in principle may be met by trade; and the poorest within any country may go hungry. To the extent that population growth exacerbates income inequalities, for example by enlarging the class of rural landless, it may make the food problem worse.

Hans Linnemann's chapter discusses Srinivasan's contribution. While agreeing with Srinivasan's main points, he suggests a number of reasons for taking a somewhat more pessimistic view. Linnemann also considers whether the protective agricultural policies of the industrial countries may contribute to the poverty in the agricultural sectors of the Third World, by depressing international food prices. Using a model discussed earlier by Srinivasan, he reports some simulations indicating that this is so.

Of course, projections spanning two decades represent only a small portion of the recent past and the likely future, but these illustrate, none the less, the capacity of the agricultural economies to adjust to changing demands and pressures. The adjustments to population growth are explored in more detail, but from a general point of view, in two chapters. The one by Rosenzweig, Binswanger, and McIntire discusses the effects of increasing population density on institutional arrangements governing exchange and ownership, with particular attention to the environment of risk and the limits of information. Presenting a more formal development of some ideas of Ester Boserup and developing many of their own, they argue that population growth and land scarcity lead in the long run to the development of labour markets, property rights in lands, landlessness, credit markets, and purchased inputs, and also bring about reduction in common property resources, livestock tenancy, and horizontally extended households.

A better-known theme in Boserup's work is the effect of population growth and density on investment, technology, and choice of technique. This theme is elaborated, particularly in the light of agricultural practices in Africa, in the chapter by Pingali and Binswanger. They show a remarkable degree of substitutability of capital and labour for land, so that in the long run returns to agricultural labour appear to decline quite slowly as population density increases. This is particularly so on wet lowlands, which are avoided at lower densities in favour of the hillsides but respond more readily to intensification.

The general ideas in the Pingali–Binswanger chapter are given an explicit econometric representation and applied to North India in a chapter by Robert Evenson. Although he discusses a specific region of a specific country, the methods and underlying argument are sufficiently general that this is perhaps best viewed as a general rather than a country study. Influences on agricultural output are categorized into short-run profit-maximizing behaviour of farmers, long-run market processes (such as investment by farmers), actions of state and federal governments (such as building of infrastructure and expenditure on research), and variables beyond the control of farmers and governments. The determinants, including population density, of each of these kinds of influence except the last are modelled and estimated. When the long-run effects of a 10 per cent increase in population are simulated in the model, it is found that the negative effects would depress per capita income by 8 per cent, but that these would be partially offset by positive Boserupian effects of 3 per cent, leaving a net decline of per capita income of 5 per cent.

The chapters reviewed so far have considered how increased population density can lead to changed institutional structures, to intensification of land use, to investment in agriculture and rural infrastructure, and to technological progress. A chapter by Mead Cain and Geoffrey McNicoll argues that the consequences of population growth will be mediated by the pre-existing institutional structure and should not be considered in abstraction from it. In particular they focus on two aspects: whether the family form is nuclear or joint, and whether the form of community organization is territorial and corporate. The nuclear family system prevalent in pre-industrial Europe controlled the formation of new households and tended to preserve intact the family holding from one generation to the next. Territoriality leads to community concern with population growth; corporateness enables the community to exert pressure on individual behaviour. The fourfold classification of family and community structures resulting from these categories should prove useful in analysing the responses of different societies to similar demographic shocks.

These chapters set out the general theoretical background, a background enriched by the studies of individual regions or countries which follow.

These more specific studies also pay particular attention to distributional issues, in addition to the general tendencies of rural productivity and output. In this regard, two broad questions are addressed: what is the role of population growth in creating or enlarging the class of the rural landless and how are its effects mediated by rural labour markets?

We may begin with the chapter by Azizur Khan, which discusses the problem of access to land, viewing the experience of a number of Asian countries: Bangladesh, India, Pakistan, China, Taiwan, Thailand, the Philippines, and South Korea. He first outlines what he calls the archetypal case: population growth leads to a growing labour force, which is less than fully absorbed into industry; the amount of land is fixed or grows only slightly, so land per worker declines, leading to reduced productivity of labour and falling real wages; at the same time, there is a disproportionately large increase in landlessness or near-landlessness. In fact, of the countries examined, only Bangladesh conforms to this archetypal case. In Thailand and the Philippines, the land in use expanded rapidly enough to prevent a decline in the land–labour ratio. In Korea and Taiwan, the industrial sectors grew rapidly enough to leave stable or declining numbers in agriculture. The expansion of rural non-agricultural employment, as in China, may also mitigate the effect of declining land–person ratios. Many contemporary Asian economies have been able to absorb additional labour in agriculture very productively, sometimes with rising real wages, as was historically true in Japan; examples are Korea, Taiwan, China, and Java. In the face of declining land–person ratios, redistributive measures can prevent a rise in landlessness, as was the case in China, Korea, and Taiwan. Increased landlessness need not lead to rural poverty; increased demand for labour, as occurred in Pakistan, can lead to rising real wages. Nor is a declining land–labour ratio necessary to generate growing landlessness, as the case of the Philippines shows. There are thus various 'escape routes' from the effects on the rural sector of rapid population growth. To what extent these escape routes can be expected to operate in the long run is not yet clear, but recent experience of countries such as China suggests that in some instances the possibilities may have been largely exhausted already.

Europe was able to mitigate the effects of its more moderate rate of natural increase in the nineteenth century through emigration, but this is widely held not to be a real possibility for the Third World today. None the less, some of the oil-rich countries of the Middle East have recruited considerable numbers of foreign workers. A conference paper subsequently withdrawn by its author, Samir Radwan,[1] discussed the impact of this recruitment on Egypt, providing an interesting example of the effects of

---

[1] Samir Radwan, Population·Growth, Agrarian Change and Labour Markets: The Case of Egypt, paper presented at the IUSSP Seminar on Population, Food, and Rural Development, New Delhi, India, December 1984.

emigration on the sending economy. In the 1960s, Egypt was viewed as the pre-eminent example of the labour surplus economy, with stagnant real wages and rapid population growth. In the 1980s, Egypt instead shows striking sectoral labour shortages, with a surplus of educated workers. The labour force in agriculture has declined by 10 per cent from 1971 to 1979, while agricultural real wages have risen by 235 per cent over the same period, led by construction sector, where wages rose by about 500 per cent over this period. The rise in wages and reduction in argicultural employment were both largely due to the emigration of about one million workers (roughly 10 per cent of the labour force) since 1973. Not only did this considerably reduce the potential labour force below trend, but the foreign remittances it generated helped fuel the spectacular growth of the construction industry and strengthened the foreign-exchange position of the government. A dramatic increase in government and military manpower, unrelated to the emigration, also had an important effect. Although it does appear the emigration had strong positive effects on the economy in general and unskilled and semi-skilled wages in particular, the longer-run outlook is precarious, and the possible return of the migrants after the end of the construction phase in the receiving countries could be disastrous.

International emigration has been one of the historic safety-valves for rapid population growth; an internal open frontier of unsettled lands has been another. The role of the frontier in a contemporary developing country is illustrated by George Martine in his chapter on Brazil. From this point of view, the Brazilian experience is disappointing. The enormous Amazon region absorbed only about half a million rural migrants in the decade of the 1970s, despite government plans to settle ten times that number. Furthermore, it appears that the life cycle of frontier areas, in terms of the time lapsed between intensive immigration, stagnation, and out-migration, has shortened from about 30–35 years for the earlier-settled frontier region of Paraná, to about 20–25 years for the next-settled central western and Maranhão area, to perhaps only 10–15 years for the Amazon. This progressive lack of success in settling the frontier regions apparently reflects in part the greater distances to markets exacerbated by rising fuel costs, lower soil fertility, and natural obstacles. However, government policies have also contributed to the problem by granting easier access to the frontier lands for those who already had large holdings or other sources of wealth, and by as usual denying access to the credit necessary to purchase the complementary inputs for the poor who had no collateral. Generally, more intensive cultivation of the already settled lands closer to the urban centres appeared more attractive. The outcome might have been quite different had not industrialization been proceeding so rapidly.

For India, Pranab Bardhan shows that higher population densities are associated with higher proportions of the labour force engaged in wage farm labour, in a large cross-area sample. A complex pattern of interaction with agro-climatic and institutional factors makes this association difficult to interpret, however. In a more homogeneous sample from rural West Bengal, it is shown that size of area cultivated by a household is inversely related to the probability that a woman in the household will be doing wage labour; however, this would not necessarily translate into an aggregate-level association between density and wage labour.

Graham Chipande discusses the situation in Malawi, where agricultural production is divided between smallholdings and large estates. The government has sought to encourage the smallholder sector through provision of credit, extension services, and so on. However, demographic pressures have resulted in cultivation of inferior marginal lands and a substantially reduced average cropped area per household. This has led to emigration of male labour to the more rapidly growing estate sector and to non-agricultural and foreign employment. At the same time, the more progressive households in the smallholder sector adopt the new technologies and end up hiring labour from the less progressive households, leading to income inequalities within the smallholder sector. On the whole, it appears that demographic pressures have inhibited the development of the smallholder sector, although because of the tribal ownership of the land these pressures do not lead to landlessness.

Part I

An Analytic Review of Projections of Food Output in Relation to Population

1 Population Growth and Food

An Assessment of Issues, Models, and Projections

T. N. SRINIVASAN

Department of Economics, Yale University

The impact of population size on the demand for and supply of food has long attracted the attention of economists and demographers. Most of the contemporary developed countries are currently experiencing only a slow growth, if any, in the size of their populations. Given their high real income levels, their demand for food is unlikely to grow rapidly, and so their impact on the world food economy is more significant through food supplies and exports, although the imports of a few developed countries such as the USSR on international trade in food and feed-grains could be important. For example, while the cereal imports of industrialized market economies were virtually unchanged at about 65.5 and 66.1 million tonnes respectively in 1974 and 1982, the imports of the USSR rose from 7.8 to 40.1 million tonnes and the imports of less developed countries rose from 64.2 to 95.6 million tonnes (World Bank 1984, 228, table 6). But it is fair to say that population growth is unlikely to influence the import demand for grains by the USSR. Because of this, almost all recent analyses of the food-population nexus have focused on the developing countries.

Several channels of influence in each direction can be distinguished in the relationship between population and food. First, population growth, hence the size of the future population, obviously affects the demand for food. With growth of income kept unchanged, an exogenous increase in the rate of growth of population will imply a slower growth of income per head and a slower growth of its per capita food demand. As long as the elasticity of per capita food demand with respect to income is less than unity, however, the rate of growth of total demand for food will increase with an increase in the

I wish to thank Paul Demeny, Richard Easterlin, Geoff Greene, Gale Johnson, Allen Kelley, Ronald Lee, Hans Linnemann, Eva Mueller, and Julian Simon for their comments on earlier drafts. I thank Kirit Parikh, leader of the Food and Agriculture Project at the International Institute of Applied Systems Analysis, Laxenburg, Austria for letting me use some of the preliminary results from the project. Note that this chapter appeared in substantially the same form in Johnson and Lee 1987. However, the paper was originally prepared for the IUSSP conference on which this volume is based.

rate of growth of population. Second, to the extent demand elasticities differ across socio-economic groups, changes in income distribution will have an impact on food demand even if aggregate income growth is kept constant. And the process of population growth itself can alter income distribution. Finally, population growth can affect food supplies in several ways: by changing potential labour force size and quality, by changing the availabilities (per worker) of other inputs such as land through changes in the size distribution of farms and the extent of land fragmentation, by influencing the technology of cultivation (Boserup 1965, 1981; Simon 1981), and by influencing the environment through changes in the process of soil erosion and degradation, thereby affecting yields.

The well-known channel of influence in the opposite direction, from food to population, is the Malthusian one. An increase in the availability of food over subsistence needs increases the rate of natural increase of population in the simple Malthusian model. In modern versions of economic determinants of fertility, opportunities of productive employment in household enterprises including the family farm as well as income-earning opportunities (particularly for women) outside the household enterprise affect household fertility decisions. It is clear that as economic development gathers momentum, not only will agricultural and food output rise but institutional arrangements such as particular forms of land tenure and tenancy will change. By altering the set of opportunities available for income generation in and out of farming and opening up new avenues of employment in agricultural processing and in the supply of agricultural inputs, agricultural development will influence fertility and population growth. This influence will be particularly pronounced in most of the large and heavily populated developing countries of South Asia and China where agriculture in general and production of food in particular are the activities that employ nearly two-thirds or more of the labour force.

This chapter is limited in scope, ignoring some general issues relating to population growth and economic development which have been covered by McNicoll (1984) and by the World Bank (1984) (see also Simon and Gobin 1980). First, taking the projected increases in population and per capita income until the year 2000 as given, it examines whether the global food economy can generate enough supplies to avoid a sustained increase in the relative price of food that otherwise would have to occur to bring about a balance between supply and demand, with only a brief look at the longer-term supply–demand balance. Second, it analyses the likely impact of exogenous reduction in rate of growth of population on the food consumption and energy intake of the poor. Finally, it assesses the strengths and weaknesses of some recent models of the world food economy. In particular, a model of the Indian economy is used to assess the impact of alternative assumptions regarding the growth of Indian population until the year 2000. The models reviewed vary in their approach to modelling production and supply, whether they distinguish countries and regions as well as socio-economic groups within countries in deriving demand, whether they are partial or general in modelling market equilibrium, and whether they are static or truly dynamic.

The following section discusses studies based on the concept of population-carrying capacity: the maximum population that can be sustained indefinitely into the future. By themselves these are of limited use, being technical rather than economic analyses, with little to say concerning the process of adjustment of population to carrying capacity. This section also includes a brief discussion of recent population projections for major areas of the world. The next section provides a brief description of the projections of the Food and Agriculture Organization (FAO). A discussion of the grain–oil-seed–livestock (GOL) model underlying the food supply–demand projections of the *Global 2000 Report* by the Council on Environmental Quality to the president of the United States follows. The section after that reports on the results of some simulations with the linked system of country models under the auspices of the International Institute for Applied Systems Analysis (IIASA). The subsequent section is devoted to a discussion of other projections. More speculatively the next section takes up the feedback effects in the food–population nexus neglected altogether or inadequately addressed in the models of the previous sections. Finally, the chapter concludes with a discussion of issues for further research.

Projections of Population Size and Population-carrying Capacity

Table 1.1 reproduces data from the World Bank (1984) dealing with population change and economic development. Under the standard projection, the population of the less-developed countries (LDCs) will increase from 3.4 billion in 1982 to 4.8 billion in 2000 and 8.3 billion in 2050, representing an average rate of growth of about 2.2 percent per year up to 2000 and a little over 1 percent per year for the subsequent 50 years. The projections of the US Bureau of the Census (reproduced in the Council on Environmental Quality *Global 2000 Report*) are somewhat higher, with middle-range projections of 6.4 billion in 2000, bracketed by a high of 6.8 billion and a low of 5.9 billion. The less-developed regions were projected to have a population ranging from a low 4.6 billion to a medium-range projection of 5 billion and a high of 5.4 billion. The medium variant of the United Nations projection for the world population is 6.1 billion. Though these projections differ somewhat, they all imply substantial growth by historical standards.

It is tempting to compare the projected population, say by the year 2050, with the potential for feeding this population. The study on population-carrying capacities undertaken jointly by the FAO, the United Nations Fund for Population Activities (UNFPA), and IIASA (Higgins *et al.* 1983, Shah *et al.* 1984) provides a basis for such a comparison, though it excludes some

Table 1.1   Population Projections (in millions)

Country grouping	Standard projection			Rapid fertility decline		Rapid fertility and mortality decline	
	Mid-1982	2000	2050	2000	2050	2000	2050
Low-income countries	2,276	3,107	5,092	2,917	4,021	2,931	4,225
China	1,008	1,196	1,450	1,196	1,450	1,185	1,462
India	717	994	1,513	927	1,313	938	1,406
Bangladesh	93	157	357	136	212	139	230
Pakistan	87	140	302	120	181	122	197
Middle-income countries	1,120	1,695	3,144	1,542	2,321	1,556	2,437
Indonesia	153	212	330	197	285	198	298
Nigeria	91	169	471	143	243	147	265
Brazil	127	181	279	168	239	169	247
Mexico	73	109	182	101	155	101	160
High-income oil exporting countries	17	33	77	30	46	30	49
Industrial market economies	723	780					
United States	232	259					
Japan	118	128					
Eastern European non-market economies	384	431					
USSR	270	306					
Total*	4,520	6,046					

* Excludes countries with population less than 1 million.
Source: World Bank 1984.

major countries, such as China. The objectives of the study were 'to ascertain on the basis of land resource inventories, the potential population supporting capacities in the developing world with various levels of inputs. And, second, to compare these estimates with data on present and projected populations' (Higgins *et al.* 1983, 5). Some of the earlier attempts (reviewed by Shah *et al.* 1984) estimated potential arable land and yield per hectare in different regions of the world to arrive at an estimate of potential output in grain equivalent units, then divided by an assumed consumption level per head to obtain an estimate of population potential. These estimates depended on variations in three inputs: estimates of arable land, yield per hectare, and per capita consumption needs. The range was enormous: from a low estimate of 902 million by Pearson and Harper in 1945 to 147 billion by Clark in 1967 (Shah *et al.* 1984, 5)!

The FAO–UNFPA–IIASA study differs from the earlier studies in its use of a more disaggregated data base and superior methodology. It combines a climate map providing spatial information on temperature and moisture conditions with a soil map providing spatial data on soil texture, slope, and phase, and then divides the study area into grids of 100 km² each. In all, 14 major climates during the growing period were distinguished and the 15 most widely grown food crops were considered, including wheat, rice, maize, barley, sorghum, pearl millet, white potato, sweet potato, cassava, phaselous bean, soya bean, groundnut, sugar cane, banana/plantain, and oil palm. Three alternative levels of farm technology were postulated, varying from no change in existing cropping patterns, no use of fertilizers and pesticides, and no mechanization to optimum use of plant genetic potential along with needed fertilizers and pesticides and full mechanization.

The soil characteristic, climate, length of growing season, technology, and cropping pattern together with the requirement that production be sustainable (using appropriate fallowing requirements and soil conservation measures) determine the production potential in each soil–climate grid. These are aggregated to yield production potential for each country. Deduction of seed, feed, and wastage provides an estimate of the potential output available for human consumption for each crop. Livestock production potential was also assessed under the assumption that only grassland will be used to support herds and also under the assumption that crop residues and by-products will be used in addition (Shah *et al.* 1984, 32). Given average calorie and protein requirements based on the 1973 recommendations of an expert committee of the FAO and the World Health Organization (WHO), the projected age and sex distribution of the population of a country, and the food production available for human consumption in terms of energy and protein, the maximum population which can be supported is estimated. The results are shown in Table 1.2.

In this table, Critical countries are those which cannot meet the basic food needs of their population even if all their arable land were devoted to growing food crops. Limited countries are the ones that cannot meet these needs if part of their arable land has to be diverted to produce other food and non-food cash crops. That is, if a third of the arable land in these countries is assumed to be devoted to non-food or food crops other than the basic 15, then their projected populations by 2000 would exceed their estimated carrying capacity. Land used for other agriculture would have to be converted to production of the 15 food crops identified in the study if carrying capacity were to expanded to accommodate the population. Finally, Surplus countries are the ones that meet their food as well other non-food crop requirements. Since in many countries of the developing world population will still be growing in the year 2000, Shah *et al.* (1984) compare population-carrying capacity with the hypothetical size of stationary population. In this comparison, even with a high level of technology eleven countries cannot support the size of their stationary population, while eight countries can support their stationary population only at a high level of technology.

Table 1.2 Population-carrying capacities

Region	Level of farming technology					
	Low		Intermediate		High	
Africa						
Number						
Critical countries	29		12		4	
Limited countries	4		7		4	
Surplus countries	18		32		43	
Population-carrying capacity (millions)						
Critical countries	209	(466)	62	(110)	9	(11)
Limited countries	68	(62)	340	(258)	70	(52)
Surplus countries	977	(252)	4,087	(412)	12,789	(717)
All countries	1,254	(780)	4,489	(780)	12,868	(780)
South-west Asia						
Number						
Critical countries	14		14		11	
Limited countries	1		—		3	
Surplus countries	—		1		1	
Population-carrying capacity (millions)						
Critical countries	87	(195)	116	(195)	47	(89)
Limited countries	93	(69)	—		118	(106)
Surplus countries	—		121	(69)	159	(69)
All countries	180	(264)	237	(264)	324	(264)
South-east Asia						
Number						
Critical countries	6		2		1	
Limited countries	4		—		1	
Surplus countries	6		14		14	
Population-carrying capacity (milions)						
Critical countries	270	(341)	148	(156)		(3)
Limited countries	1,492	(1,190)	—		185	(153)
Surplus countries	702	(407)	4,210	(1,782)	6,149	(1,782)
All countries	2,464	(1,938)	4,358	(1,938)	6,334	(1,938)
Central America						
Number						
Critical countries	14		7		2	
Limited countries	2		—		1	
Surplus countries	5		14		18	
Population-carrying capacity (millions)						
Critical countries	34	(52)	17	(24)	1	(2)
Limited countries	194	(139)	—		11	(10)
Surplus countries	64	(24)	540	(191)	1,281	(203)
All countries	292	(215)	557	(215)	1,293	(215)
South America						
Number						
Surplus countries	13		13		13	
Population-carrying capacity						
Surplus countries	1,418	(393)	5,288	(393)	12,375	(393)

Table 1.2 *cont'd*

Region	Level of farming technology					
	Low		Intermediate		High	
All regions						
Number						
Critical countries	63		35		18	
Limited countries	11		7		9	
Surplus countries	42		74		89	
Population-carrying capacity (millions)						
Critical countries	600	(1,054)	343	(485)	57	(105)
Limited countries	1,847	(1,460)	340	(258)	384	(321)
Surplus countries	3,161	(1,076)	14,246	(2,847)	32,753	(3,164)
All countries	5,603	(3,590)	14,928	(3,590)	33,194	(3,590)

Note: Figures in parentheses denote the projected population by the year 2000.
Source: Shah *et al.* 1984, tables 14–18.

What inference can one draw from such studies?[1] It would appear from the FAO–UNFPA–IIASA study that there is technological capability and land resources to sustain a population as high as 33 billion (or nearly nine times the projected population of 3.6 billion in 2000) in the five regions of the developing world, excluding China. All this suggests is simply that the productivity gap between techniques in use and the associated patterns of resource allocation and known superior techniques and better allocation of resources, if exploited, is sufficient to sustain a much larger population. But such projections by themselves are not blueprints for exploiting the gap.

Indeed, there is virtually no economic analysis underlying these projections. Since farming is done by millions of individual peasants, unless it is in their private economic interest, given the prices for inputs and outputs they face and the constraints to which they are subject, they will not produce a particular set and level of crop outputs merely because it is agro-climatically and technologically feasible to produce it. In particular the investments in land, capital equipment, livestock, technical skills, and knowledge needed to attain the potential output will not be forthcoming unless the returns are adequate.

Asking whether each country or region within a country has the potential to sustain its projected year 2000 population or its eventual stationary population ignores the economic cost of such autarkic development, even if it were feasible to sustain such a population. Furthermore, fundamental ideas of comparative advantage and gains from trade between regions within

[1] Yet another study of this nature is by Bernard Gilland (1983). He obtains a maximum global out-put of 7.5 billion tons of grain equivalent by multiplying an assumed maximum yield of 5 tons of grain equivalent per hectare and an assumed availability of 1.5 billion ha of land. Gilland's assumption of a completely satisfactory diet including some meat leads him to conclude that the earth can support 7.5 billion people. A projected stationary population of roughly 11.5 billion people can be supported at a one-third lower level of consumption.

a country and between countries are absent in such analyses. At best these studies are useful in pinpointing countries where, with a technology which raises the output per unit of land to the fullest extent, even the current level of population cannot be sustained relying solely on home production. This may be taken as indicating the need for out-migration of a part of its population or for investment in production for exports to pay for food imports or some combination of both.

FAO's *Agriculture: Toward 2000*

The Food and Agriculture Organization (FAO) projections for the year 2000 in *Agriculture: Toward 2000* (FAO 1981). This study individually covers 90 developing countries, accounting for 98 per cent of the developing country population outside of China, and summarily covers 34 developed countries. It analyses the implications for agriculture of three major scenarios: a trend scenario representing a continuation of the trends since the early 1960s, a modest improvement over these trends (scenario B), and a more ambitious but still feasible rate of growth (scenario A). The agronomic and technical bases for the projections are perhaps stronger than their economic basis, but the latter is an improvement over studies of population-carrying capacity. On the other hand, while the study emphasizes that access to productive assets, particularly land and credit, has to be widely shared for successful agricultural development, it does not address existing distortions, implicitly assuming that they will continue. The medium variant of the UN population projections is common to all scenarios. The demand for agricultural products is mostly driven by exogenously specified income and population trends, except that caloric intake per capita is not allowed to *fall* in countries with declining trends and not allowed to exceed certain upper bounds in countries with rising trends. Production estimates are based on projections of land and water resources, investment, and increases in yield per hectare of land.

The results are given in Table 1.3. The study concludes that doubling of agricultural production in 20 years in the ambitious scenario A (and an 80 per cent increase in the less ambitious scenario B) depends on a tremendous transformation of agriculture in all developing countries that is no less than 'almost an agricultural revolution, involving widespread modernization in technology and techniques, and based primarily on a massive increase in inputs into agriculture (well over doubling annual investment and no less than tripling current inputs alone in scenario A) . . . and pursued with an increased awareness of the need to conserve the environment and avoid undesirable social consequences.' (FAO 1981, 57.) Yet even if this ambitious task is accomplished, the study concludes, 260 million people (390 million in scenario B) in 86 of the 90 study countries, constituting 7 per cent of their

Table 1.3   FAO projections

	1980	1990	2000
Population (millions)			
90 developing countries (included in the study)	2,259	2,906	3,630
Other developing countries (including China)	993	1,121	1,244
Developed countries	1,163	1,248	1,325
World	4,415	5,275	6,199
Population growth rates[a] (per cent per year)			
90 developing countries	2.5	2.3	
Other developing countries	1.2	1.0	
Developed countries	0.7	0.6	
World	1.6	1.7	
GDP growth rate (per cent per year)[b]			
90 developing countries			
Scenario A	6.8	7.2	
Scenario B	5.6	5.8	
Developed countries			
Scenario A	3.7	3.1	
Scenario B	3.8	3.2	
Caloric intake per capita (kilocalories per day)			
90 developing countries	2,180		
Continuation of trends		2,330	2,370
Scenario A		2,445	2,635
Scenario B		2,380	2,500
Developed countries	3,315	3,415	3,475
Production of cereals (million tonnes)			
90 developing countries	382[a]		
Continuation of trends		518	636
Scenario A		569	786
Scenario B		538	696
Developed countries	818[a]		
Continuation of trends			1,102
Scenario A			1,017
Scenario B			1,069
Net trade in cereals (million tonnes)			
90 developing countries	– 36[a]		
Continuation of trends		– 72	– 132
Scenario A		– 57	– 64
Scenario B		– 67	– 105
Other developing countries (including China)	– 16		
Scenario A		– 15	– 17
Scenario B		– 19	– 27
Developing countries	– 52		
Scenario A		– 72	– 81
Scenario B		– 86	– 132
Available land (hectares per capital)			
90 developing countries		0.29	0.25

[a] Average for 1976–9.
[b] The first and second columns refer respectively to average annual growth rates during 1980–90 and 1990–2000.
Source: FAO 1981, table 3.10 and statistical annex tables 3 and 5.

population, will be seriously undernourished in 2000. In three of the countries more than 15 per cent of the population will be seriously undernourished.

The FAO study briefly addresses the question of whether different rates of population growth compared to those assumed would materially modify the results (FAO 1983, 42). Depending on where the variations occurred, the results would be changed substantially—speeding up of population growth in already poor countries with weak agricultural and economic growth prospects could be disastrous. On the other hand, a slowing down of population growth may reduce the cereal import requirements of cereal-importing countries and the number of seriously undernourished people in the population.

The study also attempts a longer-term projection up to year 2055. With population in the developing world (including China) increasing by more than 60 per cent over its level in 2000, agricultural production will have to be nearly three times its 2000 level and nearly five times its 1980 level. Since only a few countries have reserves of arable land and water, production increase will become geographically more concentrated and the importance of international trade will be more significant. Food importers will have to rely on rapid growth of production and exports of non-agricultural products to finance their food imports. The study recognizes (without attempting to quantify) the environmental implications, particularly in terms of soil quality and erosion and water pollution, of the strategy of rapid expansion of agricultural output based on a technology intensive in the use of irrigation, chemical fertilizers, and energy. The view is put forward that there is an intertemporal trade-off between reduction of poverty of the present generation and the quality of the environment bequeathed to future generations in developing countries, and that in this trade-off, given the extreme poverty in some countries, the present generation should perhaps be favoured.

The Global 2000 Report

The *Global 2000 Report to the President* (Council on Environmental Quality 1981) candidly admits that 'collectively, the executive agencies of the government are currently incapable of presenting the President with a mutually consistent set of projections on world trend in population, resources, and the environment. While the projections . . . are probably the most internally consistent ever developed with the long-range, global models now used by the agencies, they are still plagued by inadequacies and inconsistencies.' (Council on Environmental Quality 1981, 5.) The main reason for inconsistencies is that the mutual feedback effects among population, resources, and the environment are not fully allowed for in the projections, which 'are based largely on extrapolations of past trends' (Council on Environmental Quality 1981, 4). This drawback has to be kept in mind in interpreting the results.

The GOL submodel of the report is an econometric model describing the demand, supply, and trade relating to grains, oilseed, and livestock. The exogenous variables include (regional) population and income growth rates, variables describing agricultural and trade policy, and weather. Endogenous variables include prices at which trade takes place, supply, demand, and the like. The supply equations reflect technology, (for example input–output relationships) and producer behaviour. The full model consists of three sub-models for projecting arable area, total food production and consumption, and fertilizer use. Fertilizer is a proxy for a number of variables relating to technology and its change, such as the adoption of improved crop varieties, use of pesticides, or extension of irrigation. The arable area submodel is based on reliable historical data and models total arable area as a function of GOL and non-GOL product prices, a time trend, and an estimate of maximum potential arable area. The production and consumption of non-GOL products were projected on the basis of historical relationships between GOL and non-GOL products. Consumption projections were checked against historic income and price relationship and changes in taste. The fertilizer use submodel consists of region-specific equations relating fertilizer use to total food production based on cross-sectional and time-series input–output relationships.

The purpose of the GOL model is to project world production, consumption, trade, and prices of grain, oilseed, and livestock products for 1955 and 2000. While the coverage is more extensive with respect to grains and less in respect of livestock, the model is still impressive in its commodity, regional, and price detail. The model belongs to the static equilibrium genre and thus its projections, say for 1995, are independent of its projections for any other year, say 2000. Further, the changes between any two years in variables that are exogenous to the model, such as population or per capita income, by definition are not influenced by the projections for the same two years of the endogenous variables of the model.

Three crucial assumptions underlie the projections. (*a*) No major man-made or natural shocks will occur. In particular no *climatic* change is projected, though the scenarios include optimistic and pessimistic *weather* assumptions. (*b*) Yields per hectare of land will evolve at rates comparable to their historic evolution since 1950. (*c*) Protectionist agricultural policies in Western Europe and political determination of US trade with China, Eastern Europe, and USSR will continue. Three alternative scenarios are simulated. (*a*) Alternative I is the reference or baseline scenario in which growth rates of world population and per capita income assume their median values of 1.8 per cent and 1.5 per cent respectively between 1975 and 2000. No change in weather is assumed as compared to 1950–75. Energy prices are assumed either to remain unchanged at their 1974–6 real levels or, alternatively, to double by 2000. (*b*) Alternative II is the optimistic scenario with lower population growth (1.5 per cent) and higher per capita income growth (2.4 per cent). Weather is assumed to be more favourable than

Table 1.4   Food production, consumption, and trade and price in 2000

	Alternative I	Alternative II	Alternative III
Industrialized countries			
Population growth rate (per cent per year)	0.52	0.34	0.71
Per capita income growth rate (per cent per year)	2.57	3.35	1.77
Grain production (million tonnes)	739.7–679.1	730.0	683.3
Grain consumption (million tonnes)	648.1–610.8	689.6	590.2
Grain trade (million tonnes)	+ 91.3- + 68.3	+ 42.4	+ 93.1
Food production index (1969–71 = 100)	157.0–143.7	157.1	143.5
Food consumption index (1969–71 = 100)	155.8–147.7	165.7	143.6
Centrally planned economies			
Population growth rate (per cent per year)	1.21	0.94	1.43
Per capita income growth rate (per cent per year)	2.01	3.00	1.03
Grain production (million tonnes)	722.0	746.0	691.0
Grain consumption (million tonnes)	758.5	755.4	730.0
Grain trade (million tonnes)	– 36.5	– 9.4	– 39.4
Food production index (1969–71 = 100)	174.0	179.5	166.1
Food consumption index (1969–71 = 100)	179.9	179.2	173.2
LDCs			
Population growth rate (per cent per year)	2.37	2.04	2.71
Per capita income growth rate (per cent per year)	2.01	3.00	1.03
Grain production (million tonnes)	735.0–740.6	757.0	745.3
Grain consumption (million tonnes)	789.8–772.4	790.4	799.4
Grain trade (million tonnes)	– 54.8- – 31.8	– 33.4	– 54.1
Food production index (1969–71 = 100)	244.5–247.7	268.2	246.4
Food consumption index (1969–71 = 100)	247.8–242.8	261.2	249.0
World market weighted real food prices*	145–195	130	215

*Index 1969–71 = 100.
Source: Council on Environmental Quality 1981, vol. 2, tables 6–1, 6–2, 6–7, and 6–11, pp. 78, 91–2, and 96.

during 1950–75, thereby increasing yields. Energy prices are kept unchanged relative to their 1974–6 real level. (*c*) Alternative III is the pessimistic scenario with population growth high (2.1 per cent), per capita income growth low (0.7 per cent), and unfavourable weather resulting in lower yields. Real petroleum prices more than double in this scenario relative to their 1974–6 values by 2000.

Table 1.5   Per capita grain and food consumption and daily caloric intake in 2000

	Alternative I		Alternative II		Alternative III	
	Grain (kg)	Food (index*)	Grain (kg)	Food (index*)	Grain (kg)	Food (index*)
Industrialized countries	735.0–692.4	127.7–121.2	798.3	139.1	619.2	110.0
United States	1,183.3–1,111.5	135.9–128.3	1,363.3	917.7	154.8	107.9
Western Europe	581.7–548.8	121.4–115.5	599.0	124.5	518.2	110.1
Japan	484.4–452.3	164.2–154.2	481.2	163.2	401.1	138.3
Centrally planned countries	473.9	135.8	495.1	138.4	396.5	119.0
USSR	949.9	141.4	976.4	145.2	828.4	123.7
Eastern Europe	997.6	152.1	1,012.1	154.2	920.8	141.2
China	267.8	119.1	281.8	124.7	220.0	99.9
LDCs	210.2–205.5	111.0–108.6	219.4	116.7	189.5	99.9
Latin America	282.8–278.1	127.1–125.1	306.6	136.7	243.8	110.8
North Africa/ Middle East	301.8–292.8	105.9–102.2	318.6	112.9	283.7	98.4
Other Africa LDC's	112.5–112.0	81.3–80.9	119.1	86.3	108.8	78.5
South Asia	186.7–181.0	109.2–105.8	192.4	112.5	164.9	96.4
South-east Asia	233.2–228.5	117.1–114.6	237.1	119.2	217.9	110.0
East Asia	219.5–217.3	128.7–127.3	221.3	129.7	195.5	114.2
World	352.0–343.2	117.0–114.5	373.0	126.0	302.0	104.0
Daily caloric consumption in LDCs	2,370	2,330	2,390		2,165	
Latin America	2,935	2,905	3,080		2,710	
North Africa/ Middle East	2,530	2,460	2,655		2,390	
Other Africa LDCs	1,840	1,830	1,920		1,800	
South Asia	2,180	2,130	2,230		1,985	
South-east Asia	2,400	2,365	2,425		2,310	
East Asia	2,505	2,480	2,520		2,320	

*Index 1969–71 = 100.

Source: Council on Environmental Quality 1981, vol. 2, tables 6–8 and 6–9, pp. 93–95.

The projections (Tables 1.4 and 1.5) show that even in the pessimistic third alternative consumption of food is higher by about 4 per cent in 2000 over its 1969–71 level, though grain consumption is lower by about 3 per cent from its 1969–71 level of 311 kg per capita. Under the baseline Alternative I, per capita food consumption in 2000 is higher than 1969–71 by 14.5 per cent and 17.0 per cent respectively, depending on whether real energy prices more than double between 1974–6 and 2000 or stay constant. Grain consumption is higher by 10.3 per cent and 13.2 per cent respectively under the same circumstances. Though the per capita caloric consumption in 2000 in LDCs as a whole remains unchanged at its 1969–71 level under the pessimistic third alternative and increases by 7.6 per cent to 9.5 per cent under the baseline alternative, depending on the trend in petroleum prices, there are enormous regional variations. The Sub-Saharan African countries appear to fare the worst: even in the optimistic second alternative their per capita food consumption in 2000 is lower by 13.7 per cent than their 1969–71 level and by a larger 18.7–19.1 per cent range under the baseline alternative. In South Asia and North Africa only the *pessimistic* scenario leads to a fall in food consumption per capita, by 3.6 per cent and 1.6 per cent respectively in 2000 compared to 1969–71. Thus it would appear that, except for Sub-Saharan Africa, the world has the physical and economic capacity to produce enough food to meet modest increases in demand through 2000.

The report points out that the ability to sustain this *modest* increase arises from *substantial* increases in the resources committed to food production and impressive increase in gains in resource productivity through wider adoption of improved technology and the use of land-augmenting inputs such as fertilizers and pesticides. In fact, even though arable land per capita *declines* from an average 0.39 ha in 1971–5 to 0.25 by 2000 (which happens to equal that projected by the FAO for 90 developing countries; see Table 1.3) in the reference alternative, because the use of fertilizers nearly triples from 55 kg in 1971–5 per hectare to 145 kg in 2000, food production roughly doubles over the same period. Achieving such an intensification in input usage is expensive and a formidable task. This is because increasing fertilizer use depends to a significant extent on irrigation, and creating irrigation capacity is likely to be capital-intensive. Operating the capacity created and producing the fertilizers needed are both energy-intensive. Further, managing irrigation systems efficiently is skill-intensive. Moreover, the report recognizes that the effort to increase food output through expansion of arable area, extension of irrigation, and use of chemical fertilizers and pesticides will have an impact on the environment, particularly in terms of deforestation, desertification, soil degradation (increasing salinity and erosion), chemical pollution of surface and ground waters, and so on. It concludes, however, that these problems, though serious, are manageable.

Since at the regional and country level supplies and demands do not balance, there is a substantial increase in international trade. The extent of the increase by 2000 varies from 63 per cent to 110 per cent over the 1973–5 level among alternatives. This implies that food-importing countries would have to export other commodities to finance such massive increases in their food imports. Since developing country food importers will account for 36–43 per cent of world imports in the year 2000, the financing problem is indeed a serious one. Apart from the problem of generating exportable surpluses, the task of converting them into export earnings is likely to prove daunting if the protectionist trends in the developed world intensify. Political determination of US grain trade and agricultural protectionism of Western

Europe may continue, and it would be naïve to pretend that they have no serious consequences.

The report is primarily addressed to assessing the environmental impact of global population and income trends. In its discussion of very long-term effects on climate, the report confines itself to indicating possible frequencies of extremes such as severe droughts (in those areas of the world prone to such events) if global warming or cooling were to take place. However, it does not relate trends in population income, industrialization, and suchlike to the probability of long-term cooling or warming.

The System of Models of the International Institute for Applied Systems Analysis

Unlike the static partial equilibrium GOL model of the *Global 2000 Report* the IIASA system of models is of the dynamic general equilibrium genre. It consists of a set of country models, some of which were put together by research groups within each country with a substantial degree of disaggregation and others by the research team of the Food and Agriculture Project (FAP) of IIASA. The country models were designed in such a way that they could be aggregated to a common ten sector model distinguishing nine agriculture and livestock product sectors (wheats, rice, coarse grains, bovine and ovine meats, dairy products, other animal products, protein fuels, other food, non-food argicultural products) and a single sector covering all non-agricultural activities. There are 22 aggregated country models, of which 19 are models for individual countries and 3 are for country groups consisting of the European Community (EC), the Eastern European group (including the USSR) called the Council on Mutual Economic Aid (CMEA), and a residual group consisting of the rest of the world. The 22 models are linked in a global trading system (see Parikh and Rabar 1981 for details).

Briefly stated, each aggregated model consists of a supply module and a demand module. The supply module determines the production decisions in each year on the basis of *expected prices*, which are assumed to be a function of current prices and past prices, with the emerging outputs being available for sale in the next year. The output vector thus generated represents claims to income, and agents decide on the disposition of income into consumption, savings, and investment. The difference between production and domestic use determines the net foreign trade of each country. To the extent a country receives or makes international transfers, income and domestic expenditure will differ, allowing a country to run a net deficit on its external trade equalling the value of the transfer at the prices at which trade takes place. In some country models several groups of agents (rural, urban, income classes, and so on) are distinguished, each group being endowed with its own preferences and claims to output. Several government policies are modelled: tariffs which establish a wedge between domestic and international prices, export and import quotas, buffer stock operations, domestic rationing and public food procurement and distribution systems, income transfers between agents, and so forth. The system is solved as follows: for any year of simulation, the output available for exchange is predetermined by producer decisions of the previous year. With this output, given an *exogenously specified* value of trade deficit and government policies, each country's demand—and hence its net imports corresponding to any *given* set of international prices—can be determined. If the sum of the net imports over all the countries is zero, then the given international prices are equilibrium prices. If not, prices are changed until international equilibrium is achieved. Once equilibrium prices are determined, the associated domestic prices determine the output for the next year. The output response to the expected prices is influenced also by the investment in production capacity made in the previous year, with investment demand in equilibrium being part of the aggregate demand for that year. It should also be mentioned that even though population growth is exogenously specified in all models, in some models labour force participation rates and rural–urban migration was endogenized through simple income-related behavioural equations.

The data base of the model included FAO's supply utilization accounts for about 1,000 commodities for the period 1961–76, aggregated to suit the sectoral classification of the model. The model was calibrated (that is, free parameters were chosen) to reproduce the observed prices, outputs, and trade flows of the period 1970–6 as closely as possible. The model was then run in a simulation mode for the period 1977–2000.

There are several noteworthy features of the IIASA system of models not shared by others discussed earlier. First, the system is global in nature so that, in principle, all sources of demand and supply are included. Second, agriculture is embedded in a general economic equilibrium within each model and in the global system. This is important, since the strength of economic linkages between agriculture and the rest of the economy varies between countries. In poor developing countries agriculture plays a dominant role as an employer and source of livelihood so that the fortunes of agriculture have a greater influence on the rest of the economy than vice versa. In the industrialized countries, on the other hand, the role of agriculture is minor in employment and income generation and the rest of the economy affects agriculture through demand for agricultural products and supply of agricultural inputs. Third, the model is sequential-dynamic. In principle, investment in agricultural and non-agricultural capacity, labour force growth, technical change, and other dynamic elements are explicitly incorporated in the sytem. Fourth, to the extent possible, the models attempt to include real-world policy instruments such as taxes and subsidies, import tariffs, and quotas and carefully account for their budgetary as well as other effects on the system. While the model system captures many realistic aspects of the global economic system, it pays a price (perhaps a heavy one) in the extent of its aggregation and in treating all its ten aggregate sectors of producing internationally traded goods. In particular, all of non-agriculture is aggregated into a single sector and there are no purely domestic non-traded goods and services such as transport. Nevertheless, for evaluating alternative policy scenarios in a consistent way, the models are extremely useful.

The simulation results are shown in Table 1.6. These are broadly similar to those of the *Global 2000 Report*. While the report projects a global population for the year 2000 varying from 5.9 billion in Alternative II to 6.7 billion in Alternative III with a figure of 6.4 billion for the reference Alternative I, IIASA projects a figure of 6.1 billion for its reference run. The GNP growth rates in the IIASA model are endogenous; they are exogenous in the report, with the IIASA growth rates being somewhat higher. The output of all grains in the year 2000 for the IIASA model is 1.96 billion tonnes; the range is 2.12–2.23 billion tonnes in the report. Total exports of grain in 2000 is of the order of 152 million tonnes in the IIASA model, while it ranges from 178 to 239 million tonnes in the report. It is understandable that the volume of trade is higher in the report than in the IIASA projections. The reason is that the report's model of the static partial equilibrium kind limits the extent of adjustment to changing prices. Since the country groupings in the two models are different, a direct comparison of results may not be appropriate. Still, it would appear that if we use caloric intake as an indicator of welfare, the prospects for the developing countries as a whole are somewhat better in the IIASA projections than in those of the *Global 2000 Report*.

The India model of the IIASA system is more elaborate than others in that it distinguishes five income (more precisely, household per capita real consumption expenditure) groups among rural and urban households. All households within each group have the same demand function represented by a Stone–Geary linear expenditure system. The distribution of households according to per capita household consumer expenditure is assumed to be log-normal. In this model population growth is exogenously specified and influences only the demand module. Three alternative growth paths were specified: Alternative 1 corresponds to IIASA's reference projection, Alternative 2 corresponds to the standard projections for the year 2000, and Alternative 3 corresponds to the rapid fertility decline and standard mortality decline projection of the World Bank (1984). There is a difference of 121 million persons between the projections of Alternatives 1 and 3 by 2000. The model was run in a stand-alone mode, with the time path of international prices faced by India exogenously specified to be the same as that emerging as the equilibrium path in the world-model reference run. Because population influences only per capita income and demand and not the production process, the differences between the alternatives are not large. (see Table 1.7). As is to be expected, Alternative 3, with the slowest population growth,

Table 1.6 Projections from IIASA basic linked system

| | Year | OECD | CMEA | Developing countries | | | All | World (including others) |
				Middle income	Low middle income	Low income		
Population*	1980	648	375	389	695	2,076	3,160	4,338
	1990	701	406	502	891	2,513	3,906	5,186
	2000	754	437	637	1,119	3,023	4,779	6,106
Rate of growth of population (per cent per year)	1971–80	0.79	0.90	2.63	2.58	2.12	2.28	1.84
	1980–90	0.79	0.80	2.59	2.51	1.91	2.13	1.80
	1990–2000	0.73	0.71	2.38	2.28	1.86	2.03	1.63
Rate of growth of real	1971–80	3.94	5.99	6.24	6.09	5.24	5.77	4.63
GDP (per cent per year)	1980–90	3.57	5.33	5.89	5.70	5.09	5.51	4.31
	1990–2000	3.15	4.87	5.84	5.67	4.80	5.39	4.04
Daily caloric intake	1980	3,335	3,619	2,712	2,369	2,310	2,373	2,595
(kilocalories)*	1990	3,454	3,628	2,913	2,509	2,448	2,522	2,706
	2000	3,550	3,580	3,059	2,626	2,552	2,637	2,787
Production of wheat	1980	136	127	26	21	84	131	414
(million tonnes)*	1990	181	141	31	27	112	170	519
	2000	212	156	36	32	139	207	564
Production of rice	1980	15	1	10	52	158	220	241
(million tonnes)*	1990	16	2	13	67	195	275	298
	2000	18	2	16	89	224	329	355
Production of coarse	1980	315	172	60	48	120	228	757
grain (million tonnes)*	1990	366	193	75	63	142	280	894
	2000	429	199	91	182	172	345	1,040
Production of all grains	1980	466	300	96	121	362	579	1,412
(million tonnes)*	1990	563	336	119	157	449	725	1,711
	2000	669	357	143	203	535	881	1,959
Net exports: wheat	1980	55	– 20	– 10	– 14	– 12	– 36	
(million tonnes)*	1990	84	– 22	– 17	– 23	– 25	– 65	
	2000	102	– 16	– 25	– 36	– 40	– 101	
Net exports: rice	1980	2	– 1	– 1	– 2	3	—	
(million tonnes)*	1990	2	– 1	– 3	– 7	8	– 2	
	2000	2	—	– 5	– 5	5	– 5	
Net exports: coarse	1980	40	– 13	9	– 5	– 10	– 6	
grains (million tonnes)*	1990	35	– 14	3	– 10	– 23	– 30	
	2000	48	– 12	– 8	– 19	– 39	– 67	

* Three-year average up to indicated year.
Source: FAP IIASA, private communications, June 1984. Results are preliminary and likely to change and are not to be quoted without permission of Project Leader, FAP, IIASA.

Table 1.7   Projections from India model of IIASA

	Year	Alternative 1	Alternative 2	Alternative 3
Population	1980	674	672	670
(millions)	1990	843	813	788
	2000	1,048	995	927
Rate of growth of population	1971–2000	2.249	2.057	1.808
(per cent per year)	1980–2000	2.232	1.980	1.637
	1990–2000	2.206	1.980	1.637
Rate of growth of real GDP	1971–2000	4.746	4.752	4.756
(per cent per year)	1980–2000	5.349	5.356	5.363
	1990–2000	6.077	6.090	6.100
Production of wheat	1980	33	33	33
(million tonnes)	1990	57	57	57
	2000	85	84	83
Production of rice	1980	47	47	47
(million tonnes)	1990	68	68	68
	2000	92	92	92
Production of coarse	1980	26	26	26
grains (million tonnes)	1990	32	32	32
	2000	35	34	34
Production of all grains	1980	106	106	106
(million tonnes)	1990	157	157	157
	2000	212	210	209
Daily caloric intake*				
Rural group				
1	1990	1,018 (28)	1,024 (27)	1,030 (26)
	2000	1,111 (20)	1,152 (18)	1,183 (16)
2	1990	1,958 (17)	1,959 (17)	1,961 (17)
	2000	2,125 (16)	2,159 (16)	2,184 (15)
3	1990	2,584 (19)	2,588 (19)	2,591 (19)
	2000	2,840 (20)	2,872 (20)	2,897 (20)
4	1990	2,659 (20)	2,674 (20)	2,693 (20)
	2000	2,927 (23)	2,937 (23)	2,988 (23)
5	1990	3,789 (17)	3,837 (17)	3,898 (18)
	2000	3,911 (22)	4,013 (23)	4,174 (25)
Urban group				
1	1990	1,170 (2.1)	1,172 (1.0)	1,178 (0.9)
	2000	1,173 (0.5)	1,217 (0.4)	1,261 (0.3)
2	1990	1,654 (5.7)	1,657 (5.3)	1,664 (4.9)
	2000	1,689 (3.4)	1,726 (2.9)	1,766 (2.3)
3	1990	2,029 (17)	2,039 (16)	2,052 (15)
	2000	2,040 (13)	2,073 (12)	2,115 (11)
4	1990	2,379 (35)	2,396 (35)	2,419 (34)
	2000	2,352 (34)	2,397 (33)	2,456 (32)
5	1990	3,102 (41)	3,145 (43)	3,200 (44)
	2000	3,010 (49)	3,091 (51)	3,209 (55)

* Figures in parentheses represent the population of each class in rural and urban areas as a percentage of the total rural and urban population respectively in 1990 and 2000.
Source: see Table 1.6.

leads to a minuscule speeding up in the rate of growth of real GDP. However, the impact on caloric intake and on the distribution of population among expenditure groups is more perceptible. In general, for all groups caloric intake increases as population growth decreases and the distribution of income improves with a higher proportion of the population moving to richer expenditure classes, particularly in the urban areas.

Other Projections

Since the early 1970s when the Club of Rome sponsored global systems modelling, several models have been published (see Fox and Ruttan 1983 for a brief discussion of some of the models) in which the interaction of the processes of population growth, income growth, and exhaustible resource depletion is explored with a view to identifying and characterizing a global equilibrium state that would be indefinitely sustainable. Early models such as those of Forrester (1971) and Meadows *et al.* (1972) were mechanical simulations with no empirical basis and the processes describing behaviour were devoid of economic content. Though subsequent models have remedied some of these defects and introduced economic processes, they have not been notable for the soundness of their empirical econometric basis. The projections from such models are of limited use, if not altogether meaningless, and are not reported here.

Among models which have food and agriculture as their primary concern, the Model of International Relations in Agriculture (MOIRA) (Linnemann *et al.* 1979), which is a precursor of the IIASA system of models, is notable for its attempt to incorporate behavioural economics into the analysis, with sectoral value added maximization by producers given input prices and with resources and production technology as constraints. Consumer behaviour is presented by separate demand functions for each of twelve income classes. MOIRA describes food sectors of individual countries and links these sectors by means of an equilibrium model of international trade. Two alternative income growth (high and moderate) scenarios are presented (see Table 1.8). These scenarios assume no change in policies relating to tariffs and quotas, domestic policies relating to rural–urban income parity targets, and so on. Two variants relating to rate of growth of non-agricultural GDP (a crucial variable which is exogenous) were used. In keeping population growth unchanged, lowering income growth rates lowers per capita income growth and hence shifts demand downwards. This results is lower food production and consumption, the effects being more significant for the developing countries. In tropical Africa and South Asia, food consumption per head in 2000 goes down by 4 per cent and 23 per cent respectively relative to their 1980 values in the low income growth variant, contrasted with *increases* of 48 per cent and 4 per cent in the high income growth variant, though the dramatic change in the case of tropical Africa appears somewhat peculiar.

Table 1.8   Projections of MOIRA

	High growth of non-agricultural GDP				Low growth of non-agricultural GDP	
	1990		2000		1990	2000
Food production (index 1980 = 1)	A	B	A	B		
Developing countries	1.31	1.26	1.89	1.71	1.22	1.47
Latin America	1.46	1.47	2.12	2.11	1.45	2.03
Tropical Africa	1.51	1.48	2.17	2.04	1.14	1.48
Middle East	1.33	1.31	2.29	2.15	1.18	1.81
Southern Asia	1.14	1.00	1.53	1.20	1.05	0.91
Developed countries	1.42	1.27	1.78	1.51	1.17	1.35
North America	1.47	1.24	1.74	1.40	1.13	1.25
European Community	1.35	1.28	1.73	1.50	1.17	1.35
World	1.38	1.28	1.80	1.60	1.21	1.42
Consumption per capita (index 1980 = 1)						
Developing countries	1.12	1.15	1.30	1.36	0.97	0.94
Latin America	1.21	1.28	1.46	1.63	1.04	1.08
Tropical Africa	1.26	1.26	1.48	1.48	1.04	0.96
Middle East	1.21	1.23	1.54	1.67	0.98	1.07
Southern Asia	1.04	1.00	1.12	1.04	0.92	0.77
Developed countries	1.18	1.19	1.34	1.37	1.07	1.13
North America	1.15	1.14	1.26	1.26	1.06	1.11
European Community	1.17	1.18	1.31	1.34	1.07	1.12
World	1.15	1.16	1.28	1.35	1.02	1.02
World price index of food (1965 = 1)	0.75	0.44	0.98	0.22	0.74	0.71
World population (millions)	5,237	4,318	6,146	4,722		
Developing countries	3,916	3,122	4,733	3,484		
Developed countries	1,321	1,196	1,413	1,238		
Undernourished (millions)	520	400	740	460		
As a proportion of developing countries' population	13	13	16	13		

Note: Columns A and B refer to the reference and low population growth scenarios respectively.
Source: Linnemann *et al*. 1979, and tables 8.2, appendix 10A, runs 111, 112, 211, pp. 245, 306–68.

Keeping income growth at its high value but halving population growth increases per capita growth and per capita demand. On the other hand, a lower population means lower labour availability for agricultural production. Since higher per capita demand is moderated by the lower absolute level of population, the net effect on *total* demand is downward, reflecting an income elasticity of demand less than unity. This results in lower total output and lower food prices compared to standard population growth run, but in higher per capita consumption of food everywhere except Southern

Asia, where per capita consumption is *lower* in the low population growth run. The reason for this peculiarity is that domestic food prices have to be raised substantially, compared to the high population growth run, to maintain rural–urban income parity (Linnemann *et al.* 1979, 298).

The last two sets of projections to be noted briefly here were published by Resources for the Future, Inc. (1984) and the International Food Policy Research Institute (IFPRI) (1977). The basic underlying assumptions of the former were that population would grow at an average rate of 1.75 per cent per year during 1980–90, slowing down to 1.65 per cent during 1990–5; 93 per cent of this growth would be in the LDCs. The world population would be 6.16 billion in the year 2000. Per capita real GNP would grow at an average rate of 3.5 per cent per year during 1980–2000, with the highest growth rate 5.6 per cent occurring in East Asia and the European Economic Community (EEC) (2.5 per cent), Sub-Saharan Africa (2.6–3.2 per cent), and Eastern Europe (3.1 per cent) being at the lower end. These assumptions imply that world production of cereal grows at a rate of 1.83 per cent per year during 1980–2000, with the EEC (1.12 per cent), Sub-Saharan Africa (1.70 per cent), Eastern Europe (1.16 per cent), Asian centrally planned economies (1.63 per cent), and North America (1.35 per cent) growing more slowly than the rest of the world. World trade grows substantially, with net imports of cereals rising from 131 million tons in 1978–80 to 242 million tons in 2000. Meat imports triple from 2.6 million tons to 7.4 million tons during the same period. The study concludes that 'the world possesses the potential to feed a growing population of 6.1 million people moderately better than it fed 4.3 billion in 1980' (*Resources* 1984, 19), because per capita consumption of cereals increases from 361 kg in 1978–80 to 372 kg in 2000 and meat consumption rises from 32 kg to 37 kg.

The IFPRI (1977) study is devoted to an assessment of food needs of developing countries by the year 1990. The methodology used in very simple, even simplistic: production was projected to 1990 by extending the 1960–75 trend. Consumption *targets* for 1990 were derived in alternative ways. In variant 1, average per capita food consumption of 1975 is to be provided to the 1990 population, in variant 2, given a set of income elasticities based on FAO studies, the demand arising from a low-income growth scenario (2*a*) and a high-income growth scenario (2*b*) were set as targets and finally, in variant 3 food needs to meet 110 per cent of FAO–WHO daily minimum per capita food energy requirements were set as targets. Income growth assumptions were based on World Bank projections. The medium variant of population projections of the UN was used as reference in all scenarios, while the low variant (2*a*1 and 2*b*1) was used as well in the two alternative income growth scenarios. In all, 84 developing market economies were included. The projections are reproduced in Table 1.9. These appear to fall within the range of the other projections reviewed earlier. In the low variant of population growth, the food deficit of low-income importing countries is

Table 1.9   IFPRI projections

Average growth rates (per cent per year)	Population	Food production	
	1975–90	1960–75 (cereals)	1975–90 (major staples)
Food deficit countries	2.6	2.8	2.7
Low income	2.6	2.6	2.4
Middle income	2.9	3.6	3.5
High income	2.7	2.4	2.4
Grain exporters	2.9	4.0	4.0
All 84 countries	2.7	3.0	2.9

Food consumption in 1990 (million tons)	Variant			
	1	2a	2b	3
Food deficit countries	567	627	649	654
Low income	349	385	398	427
Middle income	166	180	185	170
High income	52	62	67	56
Grain exporter	52	56	57	54
All 84 countries	619	682	706	708

Gross food deficit in 1990 (million tons)*	Variant			
	2a	2a1	2b	2b1
Food deficit countries	121	103	143	133
Low income	69	55	83	69
Middle income	21	19	25	21
High income	31	29	35	33
Grain exporter	—	—	—	—
All 84 countries	726	54	143	133

* In computing gross deficit, the exports *within* each group have not been substrated from the imports of each group.
Source: IFPRI 1977, tables 5, 6, 7, and 8, pp. 44–63.

somewhat less than in the medium variant, and that of others marginally less. Since population growth affects the two IFPRI projections being compared (2a and 2a1, 2b and 2b1) only through demand effects arising from the implied changes in per capita income, the relatively small difference observed between low and medium variants of population growth is not surprising.

The Neglected Effects

In all the projections reviewed here, population growth is exogenous. Also ignored are the effects, *if any*, of exogenous population growth on the environment through desertification and soil erosion; any consequences for climate in general (and frequency of severe droughts and floods in particular); on farm size, and increasing fragmentation of holdings resulting from subdivisions of land within families. Writers such as Lester Brown and colleagues of the Worldwatch Institute argue that 'as world population expands, the shrinking cropland area per person and the reduction in average soil depth by erosion combine to steadily reduce the per capita availability of topsoil for food production' (Brown *et al*. 1984, 189.) They conclude that 'achieving a more satisfactory balance between the world demand and supply of food requires attention in both sides of the equation. On the demand side, the success of efforts to upgrade diets may depend on an emergency program to slow population growth—On the supply side, the scarcity of new cropland, the continuing loss of top soil, the scarcity of fresh water, and diminishing returns on chemical fertilizer combine to make expanding food production progressively more difficult' (Brown *et al*. 1984, 190–1.) After pointing out that expanding food supplies may be progressively costly relative to real incomes (particularly of the poor), they conclude that 'nothing less than a wholesale reexamination and reordering of social and economic priorities—giving agriculture and family planning the emphasis they deserve—will get the world back on an economic and demographic path that will reduce hunger [rather] than increase it' (Brown *et al*. 1984, 193).

Apart from the doubtful empirical support for estimates of global soil erosion and degradation, the formulation given for the impact of population growth ignores the response of private agents to market signals as well as social action that can prevent such a grim situation from arising. As Kelley (1985) points out, essentially there are only two prima-facie plausible arguments that can be advanced in support of the hypothesis that rapid population growth will necessarily lead to disaster. The first argument holds that rapid population growth, by extending cultivation to marginal lands and intensive cultivation in intramarginal lands, will lead to a progressively increasing relative price of food because of diminishing returns to factors other than land. But this argument assumes that (*a*) reserves of arable land are nearly exhausted, (*b*) technical change that can mitigate diminishing returns will not occur, and (*c*) the benign effect of rising incomes in the rate of lowering in population growth, if there is any, will be too slow-acting relative to the malign effect of rising cost of food on the health and nutritional status of the poorer groups in the population. The second argument rests on the belief that natural resources (including the environment) are exhaustible in the sense that real marginal costs of use will eventually rise steeply and the exploitation of possibilities of substitution of relatively abundant natural resources and/or primary factors such as capital and labour are limited.

Simon (1981) has persuasively argued that empirical support for these assertions is almost non-existent. In this view, available data suggest that the real cost of food (as well as many other natural resources) has been *falling* instead of *rising*. There are still reserves of unutilized arable land in some areas of the world (particularly in South America), and the potential for increasing yields by increasing cropping intensity (that is, through multiple cropping) is far from exhausted. The potential for raising output in many parts of the developing world through adoption of *known* superior technology is yet to be realized in full measure. Also, evidence for induced innovation (Ruttan and Hayami 1984, Hayami and Kikuchi 1981) suggests that the processes of technical and institutional change themselves will be responsive to emerging scarcities. In any case, the fact that relative prices of food and many natural resources have not risen, and in many cases have fallen, suggests that the bear of diminishing returns and the bogey of resource depletion have so far been kept at bay. The externality aspects of environmental degradation and the institutions to internalize it are well known.

Some, even among those who do not foresee a rapid population growth as a problem in the long run, would still recognize that there may be an adjustment problem in the short and medium run. Ruling out adjustment through Malthusian natural checks or coercive controls over child-bearing decisions of couples, it is possible that (in the absence of non-coercive policy instruments that influence population growth) there may be no feasible path that would take the society from its initial position to the steady state, even though the long-run sustainable level (that is, steady-state) of population may be substantially larger than the current level and the level of living associated with the long-run steady state may also be much better. One reason advanced for this is the belief that rapid population growth reduces savings and investment, possibly delaying or precluding the attainment of the steady state. Despite many studies, however, empirical evidence on the negative impact of population growth on saving and investment is at best inconclusive. A more serious problem in many developing countries is that inappropriate public policy interventions have blunted and distorted the incentives of farmers to enlarge food supplies. Even in countries where substantial investment in irrigation works, development of location-specific agricultural technology (including superior crop varieties), diffusion of such technology through extension and subsidized credit, and the like have been part of public policy, the design of these policies and the management of the facilities created have been so poor and leakages endemic as to reduce their benefits and to distort their distribution among socio-economic groups in the rural population. As contrasted with the empirical support for many of the arguments about the deleterious consequences of rapid population growth, the empirical evidence on the cost of ill-conceived public policy interventions in agriculture in developing countries in Africa, Asia, and Latin America is strong and well documented.

Conclusions

It would appear from the projections of various models reviewed earlier that the demand for food likely to arise from anticipated income and population growth can be met without a substantial increase in the relative price of food. However, this conclusion must be qualified, for several reasons. Even though alternative income and population growth scenarios are analysed in almost all the models, in none of the models is population growth-endogenous and only in the IIASA model is income growth-endogenous. Endogenizing population growth is very likely to increase the chances of long-run viability of the system. Adjustment to incipient excess demands or supplies of goods and factors through relative price changes is exploited to the greatest extent in the IIASA system of models, but to lesser and varying degrees in other models. The process of technical change is very crudely modelled, if at all. In particular it is independent of population growth so that the possibility that rate of technical progress is augmented by population growth is ruled out. Except for the IIASA system and the MOIRA models all are static, but even in the former the modelling of the process of investment in capacity creation appears to be rudimentary, and future returns from investing in alternative activities do not appear to influence the pattern of investment.

There are several possible channels of influence of population growth on the production capacity of the economy in general and food and agriculture in particular, not all of which have been taken into account in the projections reviewed. One of the more important among these is the process of shift of labour away from agriculture. In most of the presently developed countries this proportion is less than 10 per cent, and in some of the developing countries it has fallen substantially since the Second World War. In India, however, it has hardly changed in over one hundred years from about 70 per cent, even though the share of agriculture in gross domestic product has steeply declined. The situation in Bangladesh is no better. The proportion of China's labour force employed in agriculture is only marginally less than that of India's, according to the World Bank (1984, table 21, 258). If a virtually unchanging proportion of a rapidly growing (because of population growth) labour force is employed in agriculture while at the same time the share of domestic products originating in agriculture falls, income disparities between agricultural and non-agricultural workers will widen, in the absence of massive transfers. But the failure to reduce the pressure on agriculture would seem to emerge not from rapid population growth *per se* but from the strategy of industrialization that raised capital intensity of production outside agriculture, thereby limiting the scope of expanding non-agricultural employment.

It was mentioned earlier that the pattern of land-holdings (in terms of size distribution of farms), land tenure, and other contractual arrangements in agriculture may be influenced by population growth and technical changes. Such institutional changes may in turn affect distribution of real incomes (or 'income entitlements', as Sen (1981) put it) and access to food. Incorporating these into the formal methodology of projections is not simple, if for no other reason than that the theory of endogenous institutional change is in its infancy. Yet these changes could be far more significant than those included in the projection models.

A satisfactory quantitative answer to the question of whether generalizing the projection models by including in a suitable way several of the above-mentioned factors will improve or worsen the prospects for accommodating increasing population at a reasonable level of living obviously cannot be given without doing such an exercise. A qualitative and somewhat speculative answer can be given, however. Recognition and utilization of the proven strength of the price system in considerably reducing, if not eliminating altogether, any incipient imbalances between supplies and demands in the short run and, more important, in providing appropriate signals for directing investments so as to ensure long-run balance formulation models and policies can only enhance the viability of the system. The influence of technical change, particularly of the endogenous or induced variety, is also likely to be in the same direction. Many structural rigidities and development policies that reduce the static and dynamic efficiency of resource use are rarely reflected in the models, and thus static once-and-for-all as well as continuing dynamic gains from their removal are absent from such models. This again works towards underestimating the *potential* strength and long-run viability of the system. On the other hand, in the absence of a satisfactory theory of institutional change, it is difficult to assess even qualitatively whether such change will be orderly and distribute the burden of adjustment in proportion to the capacity to bear such burdens. For instance, demographic pressures can, though not necessarily, lead to an increase in landlessness and in unviable, fragmented land-holdings. Unless institutions and policy-makers respond by encouraging consolidation of fragmented hold-·ings, ensuring access to land through tenancy arrangement, and augmenting income-earning opportunities outside agriculture, the extent of poverty and food insecurity among the poor may increase and eventually threaten political stability. In such a situation, it is hard to say whether easing demographic pressures will merely postpone the day of political reckoning or provide an extended period for institutions to respond positively.

Normal fluctuations in food supply, whether they relate to output or to terms of trade, have to be addressed by other means, and population growth has little to do with them. How the available food is distributed among the population will depend on the institutional arrangements relating to production and exchange. For instance, in a market economy an individual has to have enough purchasing power through 'income entitlements' to be able to afford a diet above starvation level. The nature of transportation, storage, and distribution networks are also important. As tragic events in Ethiopia have shown, in the absence of such a network the food shipped by the rest of the world will not reach the starving. These elementary relationships between institutions and access to food and their implications for understanding episodes of famine are elegantly elaborated in Sen (1981). It appears from his analysis that their main cause was not shortage of food or rapid population growth but colossal *policy* failures in areas unrelated to population growth.

Once again, with famine raging in Ethiopia, some continue to assert that even though successive droughts are contributory causes, rapid population growth and its alleged consequences (of desertification, abandonment of traditional methods of cultivation in favour of others which were ecologically damaging, and so forth) are behind the tragedy and that desertification may even be responsible for the droughts. As in earlier episodes, however, policy failures particularly, in distorting incentives, may have more to do with the tragedy than slower-acting long-term ecological processes. Comparison of the experiences of Tanzania and Zimbabwe in coping with drought indeed suggests that Tanzanian policies contributed significantly to a relative lack of success. The cause of eliminating starvation and hunger in the world in a not-too-distant future will be ill served if, instead of analysing avoidable policy failures, attention of policy-makers is devoted mostly to attempts at changing an admittedly slow-acting process of interaction between population growth and the food economy. This is not to deny the modest improvements in income distribution and the extent of under-nourishment resulting from an exogenous reduction in the rate of population growth, as shown by some of the models reviewed earlier, but only to point out that the pay-off to correcting policy failures is likely to be quicker and perhaps greater. This is an inescapable conclusion that can be drawn from the remarkable rebound from what many then viewed as a crisis (whether correctly or not) in the world food economy in 1974. Since then world food production has increased by 30 per cent, outstripping population growth. M. J. Williams, the executive director of the World Food Council (established by the World Food Conference in 1974), was quoted as saying that 'After 10 years, it's quite clear that globally the world can produce enough food to feed all its population. And that assumes a yearly increase in that population.' (*New York Times*, 2 December 1984.) The only exception to this encouraging picture is Africa. According to the same report Williams attributed only a small part of Africa's food problems to drought and a larger part to the failure of many African governments to develop farm programmes that would provide incentives to small farmers.

References

Boserup, E. (1965), *The Conditions of Agricultural Growth,* Aldine, Chicago.
—— (1981), *Population and Technological Change: A Study of Long Term Trends*, University of Chicago Press, Chicago.
Brown, L. (director) (1984), *State of the World 1984: A Worldwatch Institute Report on Progress Toward a Sustainable Society*, chap. 10, W. W. Norton, New York.
Council on Environmental Quality (1981), *The Global 2000 Report to the President:*

Entering the Twenty-first Century, vols. 1 and 2, US Government Printing Office, Washington, DC.

Food and Agriculture Organization (1981), *Agriculture: Toward 2000*, FAO, Rome.

Forrester, J. (1971), *World Dynamics*, Wright–Allen Press, Cambridge, Mass.

Fox, G., and V. Ruttan (1983), 'A guide to LDC food balance projections', *European Review of Agricultural Economics* 10, 325–56.

Gilland, B. (1983), 'Considerations on World Population and Food Supply,' *Population and Development Review* 9, 203–211.

Hayami, Y., and M. Kikuchi (1981), *Asian Village Economy at the Crossroads*, Johns Hopkins University Press, Baltimore.

Higgins, G. M., A. H. Kassam, L. Naiken, G. Fischer, and M. M. Shah (1983), *Potential Population Supporting Capacities of Lands in the Developing World*, Technical Report FPA/INT/513 of Project Land Resources for Population of the Future, FAO, Rome.

International Food Policy Research Institute (1977), *Food Needs of Developing Countries: Projections of Production and Consumption in 1990*, Research Report 3, International Food Policy Research Institute, Washington, DC.

Johnson, D. C., and R. D. Lee (1987), *Population Growth and Economic Development: Issues and Evidence*, University of Wisconsin Press.

Kelley, A. (1985), 'The Population Debate: A Status Report and Revisionist Reinterpretation', *Population Trends and Public Policy* 7, 12–23.

Linnemann, H., J. De Hoogh, M. A. Keyzer, and H. D. J. Van Heemst (1979), *MOIRA: Model of International Relations in Agriculture*, North-Holland, Amsterdam.

McNicoll, G. (1984), 'Consequences of Rapid Population Growth: An Overview and Assessment', *Population and Development Review* 10(2), 177–240.

Meadows, D. H., D. L. Meadows, J. Randers, and W. W. Behrens III (1972), *The Limits to Growth*, Universe Books, New York.

Parikh, K. S., and F. Rabar (eds.) (1981), *Food for All in a Sustainable World: The IIASA Food and Agriculture Program*, International Institute for Applied Systems Analysis, Laxenburg, Austria.

Resources for the Future (1984), 'Feeding a Hungry World', *Resources* 76, 1–20.

Ruttan, V. W., and Y. Hayami (1984), 'Towards a Theory of Induced Innovation', *Journal of Development Studies* 20(4), 203–23.

Sen, A. K. (1981), *Poverty and Famines*, Oxford University Press, Oxford.

Shah, M. M., G. Fischer, G. M. Higgins, A. H. Kassam, and L. Naiken (1984), 'People, Land and Food Production: Potentials in the Developing World', mimeo, International Institute for Applied Systems Analysis, Laxenburg, Austria.

Simon, J. (1977), *The Economics of Population Growth*, Princeton University Press, Princeton, NJ.

—— (1981), *The Ultimate Resource*, Princeton University Press, Princeton, NJ.

—— and R. Gobin (1980), 'The Relationship between Population and Economic Growth in LDCs', *Research in Population Economics* 2, 215–34.

—— and A. Steinman (1981), 'Population Growth and Phelps' Technical Progress Model: Interpretation and Generalization', *Research in Population Economics* 3, 239–54.

World Bank (1982, 1983, 1984), *World Development Report*, Oxford University Press, New York.

2 Population Growth and Food
Some Comments

HANS LINNEMANN

Free University, Amsterdam

Recent modelling efforts and projections

The chapter on population and food by T. N. Srinivasan (Chapter 1 in this volume) offers an extremely useful survey of the various projections of the world food situation that have been made since the mid-seventies.[1] It is a concise and balanced presentation with many valid and useful comments on the different methodologies used, on the significance (if any) of the results obtained, and on the factors or aspects left out or under emphasized in the various studies. All this is summarized in some main conclusions. Nearly always I find myself in full agreement with Srinivasan's statements and observations; his conclusions are also mine. In discussing his chapter, I will first try to bring to the fore some more specific points on which there might be at least a difference in emphasis between Srinivasan and me. After that, I will take up one point in particular, elaborate upon it somewhat, and report in this connection on some recent research results obtained with the IIASA model structure.

Thus, to begin with, five short points on Srinivasan's chapter. First, the second section of the paper reviews some studies that tried to establish the population-carrying capacity of the world and of individual countries and regions. Srinivasan's comment is that such exercises may be of some interest but that estimates of the agronomically and technically upper levels of production are not very telling as no economic analysis is involved yet. In other words, will the investments required to attain or even approach these output levels be forthcoming? And are not the fundamental ideas of comparative advantage and gains from trade conspicuous by their absence in such analysis? I agree with these comments, but their relevance would seem to depend on the way in which such estimates are used. Data (or rather estimates) regarding maximum levels of agricultural output potentially to be realized per unit of land under ideal conditions may be used as an upper limit to which the yield level approaches asymptotically. In world models that are dis-

---

[1] A comparison of the results of the Food and Agriculture Organization's *Agriculture: Toward 2000*, the Council on Environmental Quality's *Global 2000 Report to the President*, and the MOIRA study is also made by Linnemann (1983).

aggregated countrywise, the use of this asymptote (expressed in appropriate units of measurement) may ensure that the relative (that is, between-countries) endowment with natural resources for agricultural production is properly taken care of. As the asymptote of the upper yield level influences the slope of the cost curve in agriculture, it may help to establish where comparative advantages lie and how they may change in the course of development. In this manner, estimates of maximum output levels per hectare were used in the Model of International Relations in Agriculture (MOIRA) (Linnemann *et al.* 1979, especially, chapters 2, 4.2, and 4.3).

My second point is related to the first. Srinivasan states that these upper level of production studies may be taken as indicating the need for out-migration of a part of the population in the 'Critical' or 'Limited' countries. I agree. However, as we all know, in the present-day world of sovereign nation-states that forcefully guard their national borders, migration can no longer play the role that it has played in the past history of humankind. Probably the last major event of this nature has been the European conquest and colonization of those parts of that world that are often referred to euphemistically as the 'areas of recent settlement' and that have large untapped natural resources. Nowadays, we have a world with little or no migration across borders, a world with many high fences—creating rent income for the lucky ones who happen to live on the 'good' side of one of the fences.[2] It would be enlightening, I feel, if a sound economic analysis were made of the welfare costs of these fences, the losses that are due to the fences for some, and the advantages of the fences for others. How would the world food problem look if migration were to remain possible? And could a fair compensation mechanism be designed for those that were born too late to migrate? I realize that some may find this a silly question. I wonder whether it is.

In discussing the *Global 2000 Report* projections Srinivasan notes that 'political determination of US grain trade and agricultural protectionism of Western Europe may continue, and it would be naive to pretend that they have no serious consequences'. Again, I agree; I will come back to this issue (in my own interpretation of it) in the second part of my chapter.

My fourth point is essentially a question arising out of ignorance. In a critical discussion of the resource exhaustion argument, Srinivasan remarks: 'In any case, the fact that the relative prices of food and many natural resources have not risen and in many cases have fallen suggests that the bear of diminishing returns and the bogey of resource depletion have so far been kept at bay.' And he continues: 'The externality aspects of environmental degradation and the institutions to internalize it are well known.' To begin with the latter statement: yes, the externality aspects and the internalizing institutions are well known and discussed in economic theory—but are they known and internalized in actual practice? Could the stable or even falling

---

[2] I owe this idea of the fences and their economic implications to Dr M. A. Keyzer of the Centre for World Food Studies, Amsterdam.

real price of food possibly be due, at least in part, to the fact of the opening up of the resources of the 'areas of recent settlement' referred to earlier? The leading food exporters of today—the United States, Canada, Australia, New Zealand, and even Argentina—are all countries in which the soil resources have hardly been used until fairly recently, recently certainly in terms of the development of the human race. And are we sure that these newly utilized resources are being maintained fully, and are we sure that the costs of their maintenance or renewal are internalized in effect? I simply don't know; hence the question.

My fifth point is a short one, mainly to avoid misunderstanding. In his conclusions, Srinivasan comments on the fact that in the food models reviewed the process of technical change is very crudely modelled, if at all. In particular it is independent of population growth so that the possibility that the rate of technical progress is augmented by population growth is ruled out. I fully agree that technical progress is of great importance, that it is difficult to model it, and that it may be underestimated in the models reviewed. But it would be dangerous, I think, to conclude from this that consequently future prospects are much brighter than the models and projections show, because technical progress in agriculture *is* incorporated in them, however primitively, on the basis of historical experience. It may be incorporated through estimated equations containing a time trend, it may also be incorporated in the parameters resulting from a calibration procedure, or it may be hidden in the estimated effects of the use of certain inputs in the production process such as fertilizer. Thus, I feel that Srinivasan's statements on this point—however correct in themselves—should not lead us to think that all things may turn out to be much better if only technical progress had been modelled correctly. Data of that past from which most of the model parameters are derived also reflect a situation in which both population growth and technical progress did occur.

To round off this part of my discussion of Srinivasan's chapter, let me repeat his main conclusion from the model results: It would appear from the projections of various models reviewed earlier that the demand for food likely to arise from anticipated income and population growth can be met. That statement is correct, and I fully endorse it. But notice what it does *not* say, and also what assumptions are underlying it:

(*a*) It refers to the demand for food likely to arise from anticipated income growth, that is, to effective demand. It does *not* say that this demand will be high enough to meet everyone's basic *need* for food; it does *not* say that undernutrition will be eliminated. In fact, as can be read in the chapter, hunger will certainly *not* be eliminated by the year 2000.

(*b*) In general, the model projections do not go beyond 2000. The year 2000 is only 15 years away from this writing, and world population will continue to grow after that for quite some time. What are the prospects for after

2000? Uncertainty prevails. Inevitably we will soon have to start making serious models and projections beyond 2000.

(*c*) Limiting ourselves for the moment to the period till 2000, virtually all studies reviewed emphasize that the projections imply and presuppose the realization of huge investments in agriculture, in particular in developing countries. A considerable part of these investments will have to be made in agricultural infrastructure, requiring largely government funds. And it should be recalled that infrastructural investment often has a long pay-back period, making it less attractive than investment in manufacturing industry. Will such heavy investments in agriculture actually be forthcoming?

(*d*) Even if the projected food output levels in developing countries are realized, again most if not all projections foresee a continued increase in food imports of developing countries. Will it be possible for these countries to earn the foreign currency needed to pay for these high levels of food imports? For some, yes; for others, hardly. Food aid will remain necessary, and may for a number of countries be absolutely essential.

The main conclusion of Srinivasan's chapter is, therefore, not as optimistic as it may appear to be on first sight. Hence the question arises: is it possible to think of policy changes (other than those already taken into account in some of the projections) that might lead to better results, especially in terms of reducing undernutrition? On one model experiment regarding a new (international) policy I should like to report here briefly, however tentative its results are. The experiment concerned is a particular policy run of the world food model system that has been developed by the Food and Agriculture Project (FAP) of IIASA, as described in Srinivasan's chapter.[3]

A simulation run with reduced OECD agricultural production

The notion underlying this policy run is that the development of food production in developing countries may well have been affected adversely by too low a level of food prices in the world market. Low prices in the world food market are seen to be caused by too-easy supply conditions from the side of industrialized market economies—say, the Organization for Economic Co-operation and Development (OECD) countries. Obviously, as Srinivasan rightly emphasizes, inappropriate government policies have abounded in the Third World, and governments of less-developed countries (LDCs) are to blame also for the inadequacy of domestic producers' possibilities and incentives, but these inappropriate policies in part may have been induced, and more appropriate policies discouraged or even blocked, by distorting influences from the world market.

[3] I am indebted to Dr K. Parikh, Program Leader of FAP, for allowing this run to be made. For a description of the FAP model system, see Parikh and Rabar (1981), Fischer and Frohberg (1982), and Fischer and Frohberg (1984).

The European Community (EC) offers the best example of what I have in mind, but in most other OECD countries the domestic agricultural situation is not fundamentally different. For political reasons European agriculture is heavily protected. Partly because of the particular way in which this protection is given, partly because of government policies on agricultural research and modernization, and partly also because of demand for food increasing only slowly, surpluses in many products have developed in the OECD countries in the period since World War II. Imports of competing products originating from developing countries are kept out, and surpluses are sold at subsidized prices (or given away as food aid) irrespective of further consequences. To indicate the latter market constellation, I have used the term *easy supply conditions*. This policy from the side of the OECD countries has reduced, through low world market prices and concessional sales and otherwise, the incentive for developing countries to give high priority to expanding domestic production of food rapidly. The consequences may have been attractive (at least in the short run) for the urban population in developing countries, but they have been and still are negative for the agricultural population in LDCs. And it is well to remember that poverty, with all its implications, is most widespread in rural areas.

The question then is: would a more restricted OECD supply of food on the world market stimulate agricultural production in developing countries? And if so, would it reduce hunger and undernutrition? And what would be the consequences for the non-agricultural sectors? Simulations with MOIRA of a policy along these lines showed positive results; however, such a policy would have to be combined with support measures for the urban poor, at least initially. A similar policy scenario has been simulated with the FAP model system of IIASA. This scenario is characterized by the introduction of two production-reducing measures in all OECD countries except Turkey; in each country (*a*) the area under cultivation is reduced by 25 per cent of its 1980 value, in equal quantities per year spread over a five-year period beginning in 1981, and (*b*) the input price of fertilizer is increased by 50 per cent of its 1980 level, again in equal amounts per year over the same period.[4] These changes are superimposed on any endogenous changes in the magnitudes of these variables which would result from the functioning of the model.

The impact of the two policy measures is very considerable. Obviously, the immediate impact is greatest during the years in which the output-reducing measures are being introduced, the period 1981–6. OECD production levels decrease and exports shrink. World market relative prices increase sharply till 1987, in particular those for protein feed, coarse grains, and wheat, and then gradually decline to levels that remain significantly

---

[4] It is of interest to note here that these two policy measures can also be conceived as the core element of an environmental policy in OECD agriculture. Advocates of lower fertilizer application levels and of larger reserves of uncultivated or reforested land would probably argue the ecological desirability of this kind of measure.

higher than those of the reference run. The greater part of the 1980s is in this run essentially an adaptation period; it is not the large initial price shocks but the resulting higher world market price level of the 1990s that is the desired situation. Its implications will now be discussed, in comparison to the results of a reference run simulating the situation of unchanged policies; the reference run will be referred to as the RO run and the policy run with reduced OECD output as the G run.

World agricultural production increases somewhat more slowly, over the period 1980–2000, in the G run than in the RO run. Thus, the reduction in OECD production levels is not, in the period concerned, fully compensated for by increased output elsewhere. The changes in the structure of world trade in the G run are to a large extent the immediate consequence of the policy characteristic of that run. For the cereals, the United States, Canada, and Australia remain leading exporters but see their export share reduced; in the RO run the EC is still an exporter of wheat, but it now becomes an importer. For non-OECD countries, cereal imports are lower in the G run, or exports are higher—notably for Argentina and for India. For protein feed, the United States remains the largest exporter but loses ground to Brazil in particular.

As observed above, the world market price level for agricultural products is appreciably higher in the G run than in the RO run, but not uniformly so for all products. The relative price in the world market of protein feed is more than twice as high in 2000 in the G run, and coarse grains and wheat command much higher prices in G than in RO. Dairy prices are 25 per cent higher; for the remaining products price differences are less pronounced. These differences in the price structure between the two runs and the fact that per country and per product the impact of the world market price on the corresponding domestic price may and will be different (because of different price policy equations) explain why the overall higher agricultural price level in the G run work out differently in different countries: the internal terms of trade between agriculture and non-agriculture change in the G run in favour of agriculture, but to a varying degree. The better terms of trade for agriculture lead to a more rapid increase in agricultural output in all developing countries. Variations are very wide, however: in 2000 aggregate output is higher in the G run by more than 10 per cent in Kenya, Pakistan, and Turkey, but by 1 per cent or even less only in India, Indonesia, and Brazil.

Depending on the structure of the country model and its parameter values, these changes in agricultural growth may or may not also improve the performance of the non-agricultural sector. The differences in outcome in this respect are related to the ways in which the allocation of investment is modelled for the various countries. In all developing countries, the per capita income ratio between agriculture and non-agriculture improves in favour of agriculture in the G run, with improvements in the year 2000 ranging from only 2 per cent in India to 20 per cent in Egypt and Kenya. The improved terms of trade of the agricultural sector and its higher growth rate in the G run generally reduce labour outflow from agriculture and make the agricultural–non-agricultural labour ratio in the year 2000 a few percentage points higher in the G run than in the RO run.

The policies that characterize the G run are simulated not only to judge their effects on the production and supply side, particularly in developing countries, but also to see how they would affect poverty and food consumption levels. As for poverty, it has just been stated that the income 'parity' ratio improves in the G run: per capita income in the sector with lower income levels (that is, agriculture) is somewhat higher with respect to non-agricultural incomes because of the G run policies. Hence, poverty may be reduced to a certain extent; an increased shadow price of agricultural labour also points in this direction.

As regards the average calorie consumption per capita, differences in the development of this variable over time between the two runs are due to differences in the development of per capita income (which is for virtually all developing countries more favourable in the G run) and to differences in the relative (domestic) price levels of the various food commodities (which are higher in the G run). For the 1980–90 period, the overriding effect is that of the higher food prices: for all developing countries together, the annual rate of growth of calorie consumption per capita is reduced from 0.72 per cent in the RO run to 0.60 per cent in the G run. The very sharp price increases for many of the agricultural commodities in the adaptation period of the G run are responsible for this. By the year 1990 the extremely high prices on the world market have come down to their new, longer-term, and more 'normal' level (which is nevertheless, as said before, significantly higher in the G run than in the RO run). For the period 1990–2000, with a less extreme price constellation, the results are more positive for the G run; for the 11 individual developing countries in the system, we find eight times a (sometimes slightly) higher consumption growth rate in the G run, and only three times a lower rate of growth.

The nutritional consequences of improved domestic terms of trade of food agriculture in developing countries cannot properly be assessed, however, on the basis of average per capita consumption figures. Economic models of these countries must distinguish different socio-economic groups and incorporate income distribution mechanisms. As observed earlier, food subsidies for the urban lowest-income groups may well be needed as one of the complementary measures at the domestic level to accompany an OECD policy of more restricted world market supply of food.

All in all, a comparison of the simulation runs RO and G points in the direction of the following overall conclusions. The introduction of the two policy measures, area reduction and fertilizer price increase, over a five-year period in all OECD countries initially creates very sharp price increases in the world market for most agricultural commodities, followed by a rapid return to more stable price levels that remain higher than in the case of unchanged policies. Agriculture in the OECD countries is affected strongly but not dramatically. Agricultural production in developing countries is stimulated and grows faster as a result of the OECD measures, although the aggregate supply response may not be very high in the short run. The effects on food consumption in developing countries cannot readily be assessed; initially, during the years of extremely high prices, they might well be adverse, while later on positive effects may come to dominate.

I have discussed the results of this experimental run of the IIASA model system at some length for two reasons. First, they provide an illustration of Srinivasan's short statement about the relevance of agricultural policies in the industrialized countries for the world food situation. Second, they show that poverty and undernutrition in developing countries might possibly be reduced (in contrast to projections based on unchanged policies in OECD countries) by more careful and restrained activities of the industrialized countries in the world food market. In view of the appalling conditions in which many of the world's poor live, a further analysis of policy changes along the lines indicated seems to be fully warranted.

References

Fischer, G., and K. Frohberg (1982), 'The Basic Linked System of the Food and Agriculture Program at IIASA', *Mathematical Modelling* 3, 453–66.
—— (1984), 'The Differential Impact of Trade Liberalization in Agricultural Products on Developing and Industrialized Countries', paper presented at the 4th European Congress of Agricultural Economists, 3–7 September 1984, Kiel.
Linnemann, H. (1983), 'World Food Prospects till 2000', paper presented at the 7th World Congress of the International Economic Association, 5–9 September 1983, Madrid.
—— , J. De Hoogh, M. A. Keyzer, and H. D. J. Van Heemst (1979), *MOIRA: Model of International Relations in Agriculture*, North–Holland, Amsterdam.
Parikh, K. S., and F. Rabar (eds.) (1981), *Food for All in a Sustainable World: The IIASA Food and Agriculture Program*, International Institute of Applied Systems Analysis, Laxenburg, Austria.

Part II

The Rural Response to Increasing Density: Theories, Models, Evidence

3 Population Density and Farming Systems
The Changing Locus of Innovations and Technical Change

PRABHU PINGALI, HANS P. BINSWANGER

International Rice Research Institute, The Philippines. Employment and Rural Development Division, The World Bank, Washington, DC

Rapid population growth since the turn of the century has led to an exhaustion of the land frontier in most countries around the world, causing a decline in arable land per capita. Traditional societies throughout the world have devised remarkably similar means of coping with reductions in per capita land availability. This chapter highlights the farmer-based and modern technological options available to societies for achieving growth in agricultural output through increases in land and labour productivity.

The farmer's means of coping with increasing population densities and/or increased demand for agricultural output has been an expansion in the area under cultivation. Additional land was brought under cultivation either through a reduction in fallow periods or through the cultivation of virgin land. With the exhaustion of the land, intensive cultivation of permanent fields in the frontier became the norm. Permanent cultivation systems are characterized by land investments for terracing, drainage, and irrigation, intensive manuring systems, and a change from hand cultivation to use of animal draught power. All of Europe and East Asia and most of South Asia had made this transition to permanent cultivation of land before this century, while most of Sub-Saharan Africa is still under fallow systems today. Comparing Sub-Saharan Africa with Asia illustrates the process of agricultural intensification and the associated changes in agricultural technology.

This chapter was prepared for the International Union for the Scientific Study of Population for presentation at the Seminar on Population and Rural Development at New Delhi, India, 15–18 December 1984. The authors are staff members of the International Rice Research Institute and The World Bank respectively. However, the World Bank does not accept responsibility for the views expressed herein, which are those of the authors and should not be attributed to the World Bank or to its affiliated organizations. The findings, interpretations, and conclusions are the results of research supported in part by the bank; they do not necessarily represent official policy of the bank. The designations employed and the presentation of material in this document are solely for the convenience of the reader and do not imply the expression of any opinion whatsoever on the party of the World Bank or its affiliates concerning the legal status of any country, territory, or area, or of its authorities, or concerning the delimitation of its boundaries or national affiliation.

Farmer-generated technical change is capable of sustaining slow steadily growing populations with modest increases in agricultural output. It appears, however, to be incapable of supporting rapidly rising agricultural populations and/or rapidly rising non-agricultural demand for food. It is at this stage that large-scale irrigation systems and science- and industry-based technical changes must become major means for increasing the rate of growth in agricultural output. State-supported large-scale irrigation systems have been in existence for centuries in China and Egypt, and they became prominent in India in the late nineteenth and early twentieth century. Inputs such as high-yielding seed varieties and chemical fertilizers not only increase the productivity of land but also increase the productivity of labour, leading to increases in per capita output and therefore generating surpluses that can be transferred to the non-agricultural populations. However, the transition to science- and industry-based inputs is costly, especially in terms of establishing an institutional structure and an industrial base that is capable of generating and supplying these technologies. The ability of a country to achieve rapid growth in agricultural output is therefore constrained by the size of its physical and human capital base.

The next section of this chapter discusses the determinants of the intensity of land use, emphasizing the consequences of population concentration and improvements in transport infrastructure. The subsequent section discusses the farmer-based innovations in response to agricultural intensification, and the penultimate section analyses the role of science- and industry-based innovations in achieving rapid increases in agricultural output.

Determinants of the Intensity of Land Use

Population Density

The existence of a positive correlation between the intensity of land use and population density has been shown by Boserup (1965, 1981). She argues from the premise that during the neolithic period forests covered a much larger part of the land surface than today. The replacement of forests by bush and grassland was caused by (among other things) a reduction in fallow periods due to increasing population densities: 'The invasion of forest and bush by grass is more likely to happen when an increasing population of long fallow cultivators cultivate the land with more and more frequent intervals'. (Boserup 1965, 20.) Table 3.1 presents the relationship between population density and the intensity of the agricultural system. At very sparse population densities, up to four persons per square kilometer, the prevailing form of farming is the forest fallow system. A plot of forest land is cleared and cultivated for one to two years and then allowed to lie fallow for 20–25 years. This period of fallow is sufficient to allow forest regrowth. An increase in population density results, in a reduction in the period of fallow, and even-

Table 3.1   Food supply systems in tropics

Food supply systems[a]	Farming intensity[b] (R-value)[c]	Population density[d] persons/km²	Tools used
Gathering (G)	0	0–4	n/a
Forest fallow (FF)	0–10	0–4	Axe, matchet, digging stick
Bush fallow (BF)	10–40	4–64	Axe, matchet, digging stick, hoe
Short fallow (SF)	40–80	16–64	Hoes, animal traction
Annual Cropping (AC)	80–120	64–256	Animal traction, tractors

[a] Description of food supply systems:
   G: wild plants, roots, fruits, nuts;
   FF: one or two crops followed by 15–25 years of fallow;
   BF: two or more crops followed by 8–10 years of fallow;
   SF: one or two crops followed by one or two years of fallow; also known as *grass fallow;*
   AC: one crop each year;
   Multi-cropping: two or more crops in the same field each year.
Note 1: The food supply systems are not mutually exclusive. It is quite possible for two or more of the systems to exist concurrently (e.g., cultivation in concentric rings of various lengths of fallow, as in Senegal).
[b] R = number of years of cultivation × 100/number of years of cultivation = number of years of fallow. Source: Ruthenberg 1980, 16.
[c] Source: Boserup 1981, 19, 23.
[d] The population density figures are only approximations; the exact numbers depend on location-specific soil fertility and agro-climatic conditions.

tually the forest land degenerates to bush savannah. Bush fallow is characterized by cultivation of a plot of land for two to six years followed by six to ten years of fallow. The period of fallow is too short to allow forest regrowth. Increasing population densities are associated with longer periods of continuous cultivation and shorter fallow periods. Eventually the fallow period becomes too short for anything but grass growth. The transition to grass fallow occurs at population densities of around 16–64 persons per square kilometer. Further increases in population result in the movement to annual and multi-cropping, the most intensive systems of cultivation.

This leads to the broad generalization that for given agro-climatic conditions, increases in population density will gradually move the agricultural system from forest fallow to annual cultivation and even multi-cropping. Next are discussed the reasons for population concentration and/or growth and the consequent decline in arable land per capita.

Since the turn of this century, the natural rate of population growth around the world, has substantially increased, mainly due to a sharp decline in the death rates caused by rapid advances in public health services. At the world-wide level, and at the level of a specific country, the decline in arable land per capita must be attributed primarily to this general increase in population. Within a country and within regions, however, population concentrations vary by soil fertility, altitude, and market accessibility. Following are brief discussions of these intracountry variations, using examples primarily from Sub-Saharan Africa. Table 3.2 provides the major causes and consequences of population concentration.

Soil Fertility

The marginal productivity of labour is relatively higher on more fertile soils and hence one would expect immigration from less-endowed areas, leading to reductions in cultivable areas per capita. Ada district in Ethiopia, Nyanza Province in Kenya, and the southern province of Zambia are a few examples of fertile areas that are relatively densely populated and intensively cultivated.

Altitude

High-altitude areas are similarly densely populated due to immigration from the lowlands where there is higher incidence of disease (notably malaria and sleeping sickness). Population concentrations on the Highlands in Ethiopia and Kenyan are popular examples of this phenomenon.

Transport Infrastructure and Market Access

Given suitable soil conditions, areas with better access to markets either through transport networks or through proximity to urban centres will be more intensively cultivated. Intensification occurs for two reasons:

1. Higher prices and elastic demand for exportables implies that marginal utility of effort increases, hence farmers in the region begin cultivating larger areas;
2. Higher production returns on the labour invested encourage immigration into the area from neighbouring regions with higher transport costs.

Intensive groundnut production in Senegal, maize production in Kenya and Zambia, and cotton production in Uganda have all followed the installation of the railway and have been mainly concentrated in areas close to the railway line. Similarly, agricultural production around Kano, Lagos, Nairobi, Kampala, and other urban centres is extremely intensive compared to other parts of these countries. It should be noted that agricultural intensification in response to improved market access could occur even under low population densities due to individual farmers expanding their cultivated area by including more marketed crops. Interestingly, the consequences of intensification in these circumstances do not differ from those in areas with high population densities.

Table 3.2 Causes and consequences of population concentration

Causes		Direct consequence	Implications
Natural population growth:	Improved public health and lack of emigration		Reduction in fallow periods: Movement from shifting to permanent cultivation
Soil fertility:	Immigration to capture the benefits of higher returns to labour input	Reduction in available area per capita	Mechanization: Ploughing: where agro-climatic and soil conditions make it profitable
Transport facilities:*	Immigration to capture the benefits of reduced transport costs		Transport—where markets exist for food and other crops
Urban demand:*	Immigration to capture the benefits of market proximity		Milling—in response to higher opportunity cost of time for female household members
Health:	Avoidance of malaria and tsetse fly Immigration to cooler highlands		Land investments (for soil fertility, drainage, terracing, etc.): Increase in the marginal lands brought under cultivation
Historic:	Tribal war/slave trade Immigration to inaccessible highlands		Land rights: from general use rights to specific land rights
Land laws, rights:	Restrictions on the right to open new land		

* In the case of improved transport facilities and urban demand one may observe an expansion in the area under cultivation in the absence of immigration.

Other Causes

Finally, it should be noted that inter- and intra country variation in population densities, especially in Sub-Saharan Africa, have historically been caused by tribal warfare and slave trade resulting in population concentrations in relatively inaccessible highlands. Population concentration on the high plateau of Rwanda and Burundi was a response both to the incursions of slave traders and to lowland health risks. Similar migrations from the lowlands to the Mandara Mountains in Cameroon, the Jos Plateau in Nigeria, and the Rift Valley in Kenya and Tanzania have been based on the desire for personal security. Subsequent natural population growth has made many of these areas the most densely populated parts of Africa.

Implications of Agricultural Intensification

Traditional societies around the world have devised remarkably similar means of coping with reductions in agricultural land per capita caused by population concentration and/or market access. Farmer-initiated adjustments to growing land scarcity are the following:

1. A reduction in fallow periods and the concentration of cultivation on soils most responsive to intensification, such as deep clay soils;
2. An increase in land investments, such as destumping, terracing, drainage, and irrigation;
3. An increase in labour input for more intensive manuring techniques;
4. A switch from hand hoes to animal-drawn ploughs and then to tractors.

These farmer-based innovations are discussed in detail in the next section of this chapter. Bear in mind, however, that innovativeness on the part of the farmer can accommodate only slowly growing populations. Rapid growth in food output can be achieved, however, only when farmer-initiated innovations are complemented by science- and industry-based inputs, such as high-yielding varieties, fertilizers, pesticides, and similar. The penultimate section discusses the rate and direction of science-based technical change and the growth of agricultural machinery and chemical industry in response to rising population densities.

Farmer-based Innovations in Response to Intensification

Changes in Land Use

The intensification of agricultural systems is constrained by climatic and soil factors. Table 3.3 illustrates the impact of climatic factors on the intensification of the agricultural system. For given agro-climatic conditions, the extent of intensification is conditional on the relative responsiveness of the

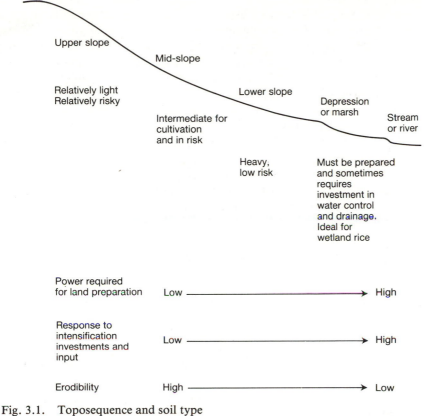

Fig. 3.1. Toposequence and soil type
Source: Pingali *et al*. 1987.

soils to inputs associated with intensive production such as land improvements, manure, and fertilizers. The responsiveness of intensification is
generally higher on soils with higher water- and nutrient-holding capacity.
This is primarily because higher water-holding capacity reduces drought
risk. Water-holding capacity is higher when the soils are deeper and the clay
content is higher. It is low on shallow sandy soils.

   Figure 3.1 presents a stylized picture of the differences in soil types across
a toposequence for given agro-climatic conditions. Soils on the upper slopes
are relatively light and easy to work by hand; tillage requirements are
minimal on these soils. The clay content and hence the heaviness of the soils
increase as one goes down the toposequence; consequently, power requirements for land preparation increase. Movement down the slope also reduces
yield risks due to increased water-retention capacity of the soils. The soils are
heaviest in the depressions and marshes at the bottom of the toposequence.

These bottomlands or *bas fonds* are often extremely hard to prepare by hand and are often impossible to cultivate in the absence of investments in water control and drainage. The extremely high labour requirements for capital investments and land preparation make the bottomlands the least favoured for cultivation under low population densities, and they are often left fallow. As population densities increase, however, the bottomlands become intensively cultivated due to the relatively higher production returns in response, to labour and land investments, especially in rice cultivation. Also, as population densities increase labour supply increases, making it possible to undertake the labour-intensive investments in irrigation, drainage, and so forth.

For instance, Grove (1961) cities the case of *fadama* (floodland) used in Northern Zaria, Nigeria. This land was not cultivated by the local farmers, who preferred the lighter soils of the mid-slopes. Seasonal migrants from the densely populated areas near Kano, however, chose to cultivate this land, as they had long been accustomed to bottomland cultivation. The farmers from Kano cultivated the *fadamas* in the dry season until the population densities in Zaria became high enough to induce local groups to undertake this type of cultivation as well. *Fadamas* in Zaria are now fully utilized and are scarce (being bought, sold, and rented).

Soil type differences across a toposequence that are characterized here could be microvariations limited to a few hundred metres or a few kilometres, or they could be microvariations where entire regions are part of one level of the toposequence. For example, the north-eastern part of Thailand can be characterized as being the upper slopes while the central plains of Thailand are the lower slopes and valley bottoms.

Preferences for cultivating different points in the toposequence also depend on the agro-climatic conditions. Table 3.3 presents soil preferences

Table 3.3   Farming intensity, agro-climates, and soil preferences

Agro-climates	Farming intensity		
	Forest and bush fallow	Grass fallow	Permanent cultivation
Arid	Lower slopes and depressions only	Lower slopes and depressions only	Lower slopes and depressions only
Semi-arid	Mid-slopes	plus Lower slopes	plus Depressions
Subhumid	Upper slopes	plus Middle and lower slopes	plus Depressions
Humid	Upper slopes	plus Middle and lower slopes	plus Depressions

by farming intensity and agro-climatic zones. Under arid conditions, lower slopes and depressions are the only lands that can be cultivated because it is only here that water-retention capacity is sufficient to sustain a crop at very low rainfall levels. This is the reason for the oasis-type intensive cultivation systems one observes in arid areas even under low population densities. Pockets of arid farming in primarily pastoral areas of Botswana are a good example of this phenomenon.

Under semi-arid conditions the mid-slopes are the first to be cultivated. As population densities increase, cultivation replaces grazing in the lower slopes and eventually in the depressions. Power sources for tillage are first used in the bottomlands generally around the time when population pressure makes these lands valuable for cultivation. The reversal of land preferences is quite dramatic. In the semi-arid zones of Africa where population density is low, the lower slopes and depressions are left for grazing and contribute only minimally to food supply. In the semi-arid zones of India, on the other hand, the depressions are intensively cultivated, usually with rice, using elaborate irrigation systems and animal traction.

Yield risks due to low water availability are not a major problem in the subhumid and humid tropics, and hence one finds cultivation starting at the upper slopes and gradually moving downwards as population pressure increases. At high population densities, the swamp and depressions become the most important land sources for food production, often associated with extremely intensive rice production. One observes such labour-intensive rice production in South and South-east Asia and could expect the same for Africa as population densities increase.

Population pressure leads to sharp reversal in preference (price) of different types of land in all but the arid zones. As population densities increase, one observes the cultivation of land which requires substantially higher labour input but which also responds more to the extra inputs. Table 3.4 provides examples of population density, agro-climates, and patterns of land use in selected African countries.

Land Investments

In the early stages of agricultural intensification (forest and early bush fallow), there are almost no investments made in land. Tree cover is cleared by felling and fire, and the stumps are left in the ground to allow quick regeneration of vegetation when the plots are returned to fallow. As a plot of land is used more permanently, the first major investment that takes place is the removal of all the tree-stumps from the fields and the setting out of well-defined plots of land on which cultivation can take place. This generally happens around the late bush fallow and early grass fallow stage of cultivation.

As mentioned previously, cultivation generally starts on the easily worked

Table 3.4   Toposequence and land use in selected African countries

Location:	Gambia	Ukara Island, Tanzania	Sukumaland, Tanzania	Mazabuka, Zambia	Zambesi Basin, Zambia	Ngamiland, Botswana
Agro-climatic zone:	Subhumid	Subhumid	Subhumid	Semi-arid	Semi-arid	Arid
Population density:	Low	Very high	High	Moderate	Very low	Very low
Patterns of Land Use:						
Upper slopes	Forest	Grazing	Grazing in wet season	Grazing	Forests	Grazing
Mid-slopes	Rain-fed groundnut, millet, cotton	Rain-fed millet, manioc	Cassava, cotton, legumes	Maize plough use	Maize and root crops	Grazing
Lower slopes	Palm forest	Legumes	Sorghum, maize	Maize plough use	Maize and taro	Millet and sorghum
Depressions	Flooded rice	Irrigated rice, sorghum, sweet potatoes, and vegetables	Irrigated rice, sweet potatoes plough use	Uncultivated	Uncultivated	Sorghum and maize plough use
References	Ruthenberg (1980)	Ludwig (1968)	von Rotenham (1971)	Trapnell and Clothier (1937)	Trapnell and Clothier (1937)	Schapera (1943)

soils of the mid-slopes. These soils are also the most susceptible to erosion as the farming system intensifies. Accordingly, systems of land use in Africa developed protective devices against erosion as population densities increase, such as ridging and tie-ridging, silt traps, and elaborate systems of stone-walled terraces. These protective land investments were already in use in the more densely populated parts of Sub-Saharan Africa prior to the colonial period (Allan 1965, 386). The hilltop refuges provided several historic examples of terrace cultivation in Africa: for instance Jos Plateau, Nigeria; Mandara Mountains, Cameroon; Kikuyu Highlands, Kenya; Mt Kilimanjaro, Tanzania; Kigezi District, Uganda; and Rwanda-Burundi (Gleave and White 1969, Morgan 1969, and Okigbo 1977).

Anti-erosion investments in land are becoming increasingly common in the more recently intensified areas of Africa. Machakos District of Kenya, for example, was a site of increased migration from the highlands, and between 1955 and 1965 the farmers in the district almost universally accepted the practice of bench-terracing the mid-slopes that are very intensively cultivated (Ahn 1977).

As population densities increase, cultivation moves from the mid-slopes to the hard-to-work soils of the lower slopes and depressions. This movement to the valley bottoms creates a need for drainage, without which the heavy waterlogged soils cannot be brought under cultivation. The draining operation is extremely labour-intensive and is generally avoided until population pressure makes the cultivation of this land remunerative. The use of the valley bottoms for rice cultivation, which is very common and a very important source of food supply in South and South-east Asia, is rare in tropical Africa. Floodland cultivation of rice in Guinea, Sierra Leone, the Senegal and Niger valleys, and the basin of Lake Victoria has been increasing, and one would expect this trend to continue throughout Africa. In Sukumaland, Tanzania, for instance, the floodplain land which forty years ago was left for grazing is now completely cultivated with rice, and the demand for this land is extremely high (Rounce 1949).

In Asia, small-scale irrigation and water-control techniques that reduce water stress or allow dry season cultivation are very common. In semi-arid India, the gently rolling hills are intensively used for rain-fed crops, the run-off being stored in tanks and used for irrigated wet rice cultivation in the valley bottoms. While some of these tank systems have been in operation for hundreds of years, the majority of the investment in these systems was made in the late nineteenth and early twentieth century. Since the 1950s, tank irrigation has been surpassed by investment in wells for cultivating a second crop on the mid-slopes; water is drawn from the wells with the help of electric or diesel pumps (Engelhardt 1984). The ultimate in water-control structures is seen in the meticulously terraced hillsides of Java and the Philippines, where in each rice field the required depth of the water is stored and the excess drained into the field immediately below (Ruthenberg 1980).

As the land frontier becomes exhausted, farmer-initiated irrigation systems have to be complemented by state-supported large-scale irrigation systems for expanding cultivation on to marginal lands and increasing the intensity of cultivation on currently cultivated land. Large-scale irrigation systems are ancient in Egypt, China, and Japan, and they have become very important in India, Korea, and Taiwan in recent decades. It is important to note that the building of such large-scale systems is induced by high population density when adequate labour supply is available and the demand for expanding cultivated area through irrigation is high. The failure of large-scale irrigation systems in Sub-Saharan Africa can be attributed to extreme labour scarcity and the lack of demand for expanding cultivated area. The Office du Niger scheme in Mali is a case in point. The 50,000 hectares that were actually developed by 1964 fell far short of the initial target of several hundred thousand hectares; even in this area, the density of settlement is insufficient to yield an output that would meet all costs of both the settlers and the management of the scheme, provide the settlers with good livelihood, and earn some return on the large amount of capital invested (de Wilde 1967, 288).

Development of Organic Fertilizer Use

Under forest and bush fallow cultivation, long-term soil fertility is maintained by periodic fallowing of land. Renewed vegetative growth on fallowed land helps to return fresh organic matter to the top soil and therefore recharges it with nutrient supplies. Also, when fire is used for clearing vegetation prior to cultivation the burnt ashes return to the soil the nutrients taken up by trees and bush cover. This closed cycle of nutrient supply is disrupted when long (forest or bush) fallow periods are replaced by short (grass) fallows.

The nutrient supply to the soil under grass fallow declines because grass cover cannot return the same amount of nutrients to the soil as tree and bush cover. Accordingly, at this stage, the farmer starts complementing fallow periods with additional organic wastes from the household, mainly in the form of vegetative waste and dung from cattle and livestock. At first, these fertilization techniques are fairly rudimentary, often involving no more than a periodic transport of household refuse to the plots to be cultivated. Sometimes, as in the case of the farmers on the Mandara mountains and Ethiopian Highlands, the dwellings are situated at a high point so that the refuse washes down to the fields below. In the lower-rainfall zones, where vegetative cover is lower, more labour input may be required to augment the supplies from the household. The Bemba in Zambia, for instance, used to cut branches from surrounding trees and carry them on to the plot of land to be cultivated and burn the pile of branches to provide nutrients for the plot (Chitemene techniques). Richards (1961), reported that branches were cut from an area up to six times as large as that to be planted.

As farming intensities increase, more labour-intensive fertilizing techniques such as composting and then manuring evolved. The Fipa of Tanzania collect fallen and cut vegetation and bury it in mounds. Beans, manioc, cowpeas, and suchlike are planted on these mounds, which are rich in nutrients due to rotting vegetation. This system of composting is common in Tanzania, Zambia, and Zaire (Miracle 1967).

The use of animal manure is common in most of the densely settled intensively cultivated pockets of Sub-Saharan Africa. Farmers in the hill refuges, mentioned earlier, have for generations used manure on their terraced fields. The inhabitants of the very densely populated Ukara Island in Lake Victoria labouriously collected three tons of manure per year from each cow, bull, or steer and transported it to the fields by head-loads (Allan 1965, 201). The use of manure is also characteristic of the densely populated parts of Northern Nigeria. In the villages of Katsina province, livestock are kept tethered in the compound and the manure is collected in heaps. Household refuse and ashes are added to the heaps. Those farmers with large holdings usually supplement their supplies through an active manure market. In the villages near Kano city, farm manure was complemented by night-soil transported from the city (Gleave and White 1969, 284). In areas where livestock herding has traditionally been separate from farming, one tends to observe contracts between herders and farmers as farming intensities increase, the typical case being a farmer inviting a herder to graze his stock on the fallow land and thus benefit from the cattle droppings. Toulmin (1983) describes such contracts in central Mali. Farmer–herder contracts are also common in India.

Finally, one observes the incorporation of legumes into a crop-rotation cycle as 'green manure'. Green manuring, along with other fertility-restoring measures, is common practice in several parts of India and China. The use of cowpeas in the rotation cycle is becoming increasingly common among the permanently cultivated areas of Africa. All of the afore-mentioned manuring techniques were (and continue to be) very important in China, Japan, and most of Europe.

As agriculture intensifies, a drop in soil fertility leads to a sharp decline in yields, which can only be reversed by more labour-intensive fertility-restoring techniques. Pingali and Binswanger (1984), using data from 52 specific locations in Africa, Asia, and Latin America show a significant positive association between manure use and farming intensity. It is important to note that farmers at lower agricultural intensities are already familiar with the more evolved manuring techniques, because many of them use these techniques on their garden plots. The reason they do not use these techniques on all their fields is that there are other alternatives, such as fallowing, which require much lower labour input than intensive manure production. It is only when land pressure makes it inevitable that they resort to general use of these techniques.

The Evolution of Tool Systems

The transition from digging sticks and hand hoes to the plough is closely correlated with the intensity of farming. The simplest form of agricultural tool, the digging stick, is most useful in the very extensive forest and bush fallow systems where no land preparation is required. But as the bush cover begins to recede, the ground needs to be loosened before sowing, and at this stage hand hoes replace digging sticks. Hand hoes are used for land preparation and weeding in the latter stages of bush fallow, grass fallow, and even in some instances of annual cultivation. But land preparation using the hoe becomes extremely labour-intensive and tedious by the grass fallow stage. This is especially true because of the persistence of grass weeds. 'The use of a plow for land preparation becomes indispensable at this stage' (Boserup 1965, 24). A switch to the animal-drawn plough during grass fallow results in a substantial reduction in the amount of labour input required for land preparation. The net benefits of switching from the hoe to the plough are conditional on soil types and topography; they are lower for sandy soils and for hilly terrain. Figure 3.2 illustrates the evolution from hand hoes to animal-drawn ploughs. This graph compares the labour costs under hand- and animal-powered cultivation systems and shows the point at which animal traction is the dominant technology.

The overhead labour costs in the transition from hand to animal power are the costs of (*a*) training animals, (*b*) destumping and levelling the fields, and (*c*) feeding and maintaining the animals on a year-round basis. The cost of training the animals is independent of the intensity of farming. The cost of destumping is extremely high under forest and early bush fallow system due to the high density of stumps per unit area and due to a highly developed root network that is difficult to remove. As the length of fallow decreases, the costs of destumping decline because of reduced tree and root density, and are minimal by the grass fallow stage. The costs of feeding and taking care of draught animals are also very high during forest and early bush fallow, primarily due to the lack of grazing land and the prevalence of diseases such as trypanasomiasis. As the fallow becomes grassy, grazing land becomes prevalent and so does animal ownership; hence the cost of maintaining draught animals declines. By the annual cultivation stage, however, grazing land becomes a limiting factor necessitating the production of fodder crops, which in turn lead to an increase in the cost feeding and maintaining draught animals. The total cost of using draught animals for land preparation, early season weeding, and manuring is given by the curve T_p (see Figure 3.2).

The labour costs for cultivation using hand tools rise rapidly as farming intensity increases. This is mainly due to the increased effort required for

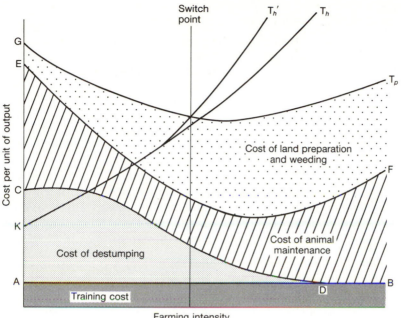

T_p = Total labour costs for land preparation, early season weeding, and manuring, using animal traction

T_h = Labour costs for land preparation and early season weeding, using the hand hoe

T_h' = T_h plus labour costs for maintaining soil fertility without manure from draught animals

Switch point = Farming intensity at which animal traction becomes the dominant technology

Fig. 3.2.   A comparison of labour costs under hand- and animal-powered cultivation
Source: Pingali *et al.* 1987.

land preparation, weeding, and maintaining soil fertility. T_h shows total labour costs using hand hoes for land preparation and weeding, while T_h' adds the cost of maintaining soil fertility. The shape of the T_h' curve depends on (*a*) the ease of producing compost, (*b*) the rate of decay of organic matter, and (*c*) the cost of chemical fertilizer. In humid and subhumid areas, it is easier to produce compost and manure than it is in semi-arid and arid areas, due to an abundance of natural vegetation; hence the labour costs involved in the production of manure are lower and the T_h' curve is flatter. In hot tropical areas the very high temperatures cause the organic matter to decay at a faster rate relative to the more temperate highlands, and hence require additional compost and manure inputs, making the T_h' curve steeper. The cheaper that chemical fertilizers are, the flatter the T_h' curve becomes due to the substitution of fertilizers for labour-intensive manure production.

Animal-drawn ploughs are the dominant technology at the point where the costs of hand cultivation exceed the costs of transition to animal power. This switch point is shown in Figure 3.2, which illustrates the following conclusions:

1. The transition to animal-drawn ploughs would not be cost-effective in forest and bush fallow systems due to the very high overhead labour costs required for destumping and animal maintenances.
2. There is a distinct point in the evolution of agricultural systems where animal draught power becomes economically feasible.
3. This dominance point is conditional on soil types and soil fertility: the transition would occur sooner for hard-to-work soils (clays) and for soils which require high labour input for maintaining soil fertility.

The complementarity between animal traction and manure use implied in the third conclusion is explained by the inverse relationship between farming intensity and soil fertility and by the increase in availability of manure when draught animals are introduced into the farming system. Fertilizers tend to substitute for manure around the annual and multi-cropping stage.

The transition from animal ploughs to tractors is explained better in the context of choice of techniques rather than in terms of the evolution of farming systems. Factors that determine this transition are capital availability, economic efficiency of tractor use, labour costs, and peak-season labour scarcity. Tractors generally emerge as feasible alternatives to animal-drawn ploughs at the stage permanent cultivation of land. Land preparation and transport are usually the first operations for which tractors are used.

A very detailed analysis on the evolution of tool systems using information from field visits to approximately 50 locations in Sub-Saharan Africa is provided in Pingali, Bigot, and Binswanger (1985). This study provides empirical evidence on the positive relationship between farming systems and tools use, and it shows that the evolution from hand hoes to animal-drawn ploughs occurs at the grass fallow stage and not before. Pingali and Binswanger (1984) provide empirical evidence on the labour-saving benefits and the consequent increases in yields per person-hour through a change to animal-drawn plough and tractors.

Science- and Industry-based Innovations

In the preceding section we showed how societies have coped with increasing population densities and/or increased market demand for food output through more intensive use of farmer-generated inputs, mainly land and labour. Such intensification can and often does increase agricultural output, but it is important to remember that the rate of growth in output is low in the absence of science- and industry-based inputs. Under these conditions, agricultural output may not keep up with, or only barely keep up with, the growth in population, and little surplus is available for transfer to the rest of the economy. This point becomes extremely clear when we consider that agricultural output is much of the developing world is growing at 1-1.5 per cent per annum (p.a.) while population is growing at 2-3 per cent p.a. Therefore, in order to sustain growing rural populations and to feed the urban population, agricultural output in the developing world would have to grow at an unprecedented 3-4 per cent p.a. This is an improbable outcome, because in the forty year period from 1930 to 1970 the developed world has averaged a growth in agricultural output of only 1.5-2 per cent p.a. (Binswanger and Ruttan 1978). The development of more productive mechanical, biological, and chemical technologies could help in narrowing the gap between the demand for and the supply of food in the developing countries. The remainder of this chapter concentrates on the potential role of science- and industry-generated innovations and inputs in increasing agricultural output.

Population Density and the Rate and Direction of Science-based Technical Change: The Induced Innovation Hypothesis

The history of agricultural growth of the developed world illustrates that the rate and direction of technical change are influenced by an economy's land and labour endowments, by the non-agricultural demand for labour and by conditions of demand for final agricultural products. In agriculture, the constraints imposed on development by an inelastic supply of labour may be offset by advances in mechanical technologies, while the constraints imposed by an inelastic supply of land may be offset by biological technology. This responsiveness of science-based invention and innovation to economy-wide factors has come to be known as the process of induced innovation (Binswanger and Ruttan 1978, Hayami and Ruttan 1971).

A comparison of agricultural development in the United States and Japan illustrates clearly the influence of land and labour endowments on the direction of technical change. This comparison highlights the extremes in land and labour endowments. In 1880, Japan had only 0.65 ha of land per male worker, while the United States had 25.4 ha, about 40 times as much. By 1970, this difference in land–labour ratio increased: Japan had 1.57 ha per worker while the United States had 160.5 ha per worker, 100 times as much. These differences in land–labour endowments are reflected in massive differences in labour-factor prices. For instance, in 1880 a worker in Japan had to work nearly 2,000 days to buy a hectare of land, while his/her US counterpart could buy land after working roughly one-tenth that time.

However, the rate of growth in agricultural output has been remarkably equal for these two countries in the ninety year period from 1880 to 1970, roughly 1.6 per cent p.a. Japan and the United States have relied on entirely different technological paths to achieve growth in output. Japan emphasized biological yield-raising technology, supported by heavy irrigation investments, along with intensive manuring techniques, while the United States emphasized mechanical technology. Careful historical and econometric enquiries by Hayami and Ruttan (1971) and Binswanger and Ruttan (1978) substantiate this conclusion.

The induced innovation literature has several implications for the rate and direction of technical change in the agriculture of developing countries. The obvious case is countries in which population pressure increases against land resources, making land increasingly scarce and expensive relative to labour. In these situations, the development of biological and chemical technologies is the most efficient way to promote agricultural growth. Technical change in agriculture in India and the Philippines are cases in point (Hayami and Ruttan 1985). Up until the end of the 1950s, growth in agricultural output was brought about primarily by expansion of the cultivated area in response to increased world demand for export crops and domestic demand for food crops. With the rapid growth of population after World War II, the supply of unexploited land became progressively exhausted. Towards the end of the 1950s, expansion of cultivated land stagnated, but the number of workers in agriculture continued to grow, leading to a decline in cultivated area per worker. In the 1960s, the response to increasing land prices and falling wages, agriculture in the Philippines made a transition from the traditional growth pattern, based on an expansion of the cultivated area, to a modern pattern, based on an increase in land productivity. Increases in land productivity were achieved through expansion of the irrigation system, modern varieties of rice, fertilizers, and other chemical inputs. Taiwan went through a similar transition from area expansion to improvements in land productivity in the 1930s (Hayami and Ruttan 1985). Innovations in biological technology that led to the rapid diffusion of the 'green revolution' (seed–fertilizer technology) in South Asia after the mid-1960s were induced by changes in relative resource endowments and factor prices similar to changes that occurred in Japan and Taiwan (Binswanger and Ruttan 1978, 360). One could expect similar development in the more densely populated parts of Sub-Saharan Africa in the future, for example as is happening in the Kenyan highlands today.

US agriculture, in contrast, emphasized mainly mechanical technology in the period 1880–1970. Innovations in biological technology did not become important in US agriculture until the 1940s, well after the land frontier had been exhausted. Examples of technical change emphasizing mechanical technology in developing countries are not common. Presumably, such change is possible where labour rather than land, is the constraint on agricultural growth and where export markets make the cultivation of larger areas profitable. Expansion in the area under sugar cane production in the southern and central western regions of Brazil were closely associated with the substitution of human labour with tractor power. The rapid adoption of tractor power in these regions during the 1960s was induced by rising wages under conditions of land abundance (Sanders and Ruttan 1978). Mechanization of European farms in Sub-Saharan Africa during the colonial period was similarly induced by severe labour constraints. Europe and Japan made the transition to mechanical inputs after the wage explosion of the 1950s, which was caused by a rapid increase in the non-agricultural demand for labour. In the late 1960s and 1970s, partial mechanization of agriculture occurred under low-wage conditions—for instance in the Indian Punjab, Thailand, and the Philippines. Mechanization under low-wage conditions is selective, concentrating only on the power-intensive operations such as tillage, transport, and processing. In all these cases, mechanical inputs for power-intensive operations co-exist with human and animal power for control-intensive operations such as weeding and interculture. Binswanger (1983) shows that it pays to mechanize the power-intensive operations even under low wages, whereas control-intensive operations are mechanized only when wages are high and/or rapidly rising.

The land and labour endowments of an economy are important determinants of the *direction* of technical change. The *rate* of technical change, however, is conditional on the economy's ability to generate and/or adapt innovations to match its specific environmental and economic conditions. In other words, the rate of technical change is determined by an economy's capital base (both industrial and human capital) and on its ability to provide institutional support for rapid technical change. To be complete, therefore, any discussion of induced technical change has to consider the environmental and institutional framework in which innovation, development, and adoption of new technology takes place.

The Generation of Innovations and the Development of an Industry

The Agricultural Machinery Industry

Mechanical technology is senitive (*a*) to agro–climatic factors such as soils, terrain, and rainfall regimes, and (*b*) to economic factors such as capital availability, farm size, and materials available. Where there is a divergence either in environmental or in economic conditions, direct transfer of mechanical technology is limited. Accordingly, where land–labour endowments warrant it, one observes a great deal of invention and/or adaptation of mechanical technology to meet local conditions. In the early phases of mechanization, such work is usually done by small manufacturers or workshops in close association with farmers. This process provides direct solutions by mechanically minded individuals to problems perceived by farmers.

For instance, in 1880 there were 800 distinct models of ploughs advertised for sale in the United States. Early machinery innovation in the developing world reveals a similar reliance on small workshops and direct farmer contact. The emergence of a diversified machinery industry out of small shops in the Indian Punjab and a power tiller industry in Thailand and in the Philippines all followed similar patterns. In the early phases, small workshops have a distinct advantage over large corporations because of (a) the location specificity of the innovations and (b) the producer's ability to capture the gains of their innovative effort through sales.

The contribution of large corporations increases over time, but it continues to be most important in the area of engineering optimization. It is at this stage that the engineering staffs of corporations are most effective. For instance, it was only around the start of the twentieth century that the plough industry in the United States consolidated, with the large firms, such as John Deere, purchasing the patents and assets of small firms as they expanded.

Given this dominant role of individual initiative in the development of agricultural machinery, what are the government policy interventions appropriate for mechanization? The government should encourage small-scale innovation and adoption through (a) patent laws for the enforcement of innovator's rights; (b) testing, standardization, and information dissemination; and (c) support of agricultural engineering education and some university-based research. Finally, it should be noted that efforts to protect the domestic agricultural machinery industry through import control have not generally been successful. This is because the small innovators would no longer have access to models or to a wide range of engines from which to design locally adapted machines. For a more complete account of the historical patterns in development of the agricultural machinery industry, see Binswanger 1983.

The Agricultural Chemical Industry

Agricultural chemical innovations are generally in the form of fertilizers, pesticides, and herbicides. The demand for agricultural chemicals (with the exception of pesticides) is induced by land–labour factor prices in the same way as the demand for biological and mechanical technology. A substitution of chemical fertilizers for farm-produced animal manure and green manure would occur only when the price of fertilizers declines relative to the price of labour. Similarly, a decline in the price of herbicides relative to the price of labour leads to a substitution of herbicides for hand weeding. For instance, given the low wage rates in semi-arid India, the use of herbicides is uneconomical compared to hand weeding (Binswanger and Shetty 1977). No particular labour-saving bias is associated with the use of pesticides, however, because pest damage to outstanding crops cannot usually be prevented by hand labour. Pesticides, of course, protect the higher output obtained through the use of fertilizers and high-yielding seeds against insects and disease and can therefore be considered a complementary (insurance) input.

Unlike in the case of mechanical technology, small entrepreneurs do not play a major role in the generation of chemical innovations. This is because the innovators require special skills acquired through university training and specialized facilities which are too expensive to provide for an individual researcher. Accordingly, most research and development of agricultural chemicals is conducted by large corporations. These corporations can accrue the returns on their investment in research through the sale of the final product, which is protected by patents. Chemical innovations have to be adapted to agro-climatic differences such as soils and rainfall regimes, but here again adaptive research is more easily done by the parent corporation. The parent company may set up experimental fields in different environments as part of its sales effort. As in the case of mechanical technology, private corporations have a comparative advantage in the research, development, and production of chemical technology. Here again the role of government should be restricted to enforcing patent laws, testing and supporting university education, and basic scientific research.

Agricultural Research Institutes

Not all agricultural research can be left to individual or private-sector initiative and innovation. There are several areas of research in which incentives for private-sector research have not been adequate to induce an optimum level of investment. In these areas the general societal rate of return exceeds the private rate of return because a large share of the gains from research are captured by other firms and by consumers rather than by the innovating firm (Ruttan 1982).

The first (and most obvious) case is basic (or supporting) research in genetics, plant pathology and physiology, soil science, and so on, which has implications for the development of chemical and biological innovations. Applied research by private corporations uses the results of basic scientific enquiry without having to compensate fully the basic researcher who produced the results.

The second case is where the search for solutions is very expensive and very risky, but where once the solutions are obtained they can easily be reproduced by the users or other firms. For instance, research and development of new crop varieties is extremely complex, having to consider a wide variety of parameters ranging from agro-climates and soil types to consumer tastes. Yet once a suitable variety is developed it can be reproduced by individual farmers. Seed companies, therefore, have not been able to capture more than a small share of the gains from the development of new crop varieties. Hybrid varieties are an exception to this generalization.

Public-sector agricultural research institutes are therefore an essential part of a strategy for rapid growth in agricultural output through science-

and industry-based inputs. Public research effort in agriculture should concentrate mainly on basic research and on research leading to advances in biological technology. Public research on mechanical and chemical technology should be minimal and mainly university-based, because the private sector has greater incentive to conduct research in this area.

It is important to remember that the role of the public sector in production of agricultural technology is enhanced only when agriculture intensifies and technologies (such as high-yielding varieties) that increase the productivity of land are demanded. Agricultural experiment stations became important sources of growth in the United States only after the land frontier was exhausted. Japan's national agricultural station became prominent earlier, in 1904, when it began initiating a crop-breeding project in the face of stagnation in agricultural output. In India, much of the post-independence consolidation and realignment of the agricultural research system took place in the 1960s, coinciding with the initiation of major crop improvement programmes. Finally, in Brazil agricultural research capacity began to develop in the 1960s and 1970s as the pressure on land and the demand of land productivity in the north-east region began to increase. (All the preceding examples were obtained from Ruttan 1982.)

Conclusion

Since the 1880s, agricultural output in the developed world has grown at a rate of around 2 per cent, irrespective of the initial factor endowments or technologies used. In addition to accommodating population growth, the agricultural output growth rate has been able to accommodate the increases in final demand for food associated with rapidly rising per capita incomes. And it has been associated, by and large, with falling real food prices.

Developing countries are experiencing rates of population growth of 1–4 per cent. If, as we all desire, their per capita incomes were to grow at rates comparable to those experienced by the developed world, food demand would grow at rates of between 2 and 5 per cent. Sustained rates of growth of food supply that exceed 2 or 2.5 per cent per year are, however, unprecedented in the history of the developed world. Of course, there exists the possibility of expanding supply via trade. But for countries where a large proportion of the economy is still in agriculture, such a strategy implies truly staggering rates of non-agricultural growth. The non-agricultural growth rates must be high enough to satisfy both increased domestic non-agricultural demand and increased exports to trade for food. Because such high non-agricultural growth rates are hard to achieve, a combination of agricultural policies and programmes must raise growth rates of agricultural supply to substantially higher levels than we are accustomed to. Such a strategy needs to encourage investment and innovation on the part of all the actors involved in the process of technology generation, investment, and production.

We have seen the truly impressive role which farmer innovation and investment has played historically, even in continents not normally associated with a dynamic agriculture. Research on the recent experience of developed countries has perhaps not sufficiently emphasized that farmer innovation and investment continue to have major importance in interaction with science- and industry-based innovations. Indeed, the increasing complexity of science- and industry-based innovation places an ever higher burden on farmers for screening and adapting technology, and it is not, therefore, surprising to find that human capital is becoming an ever more important input into the agricultural production process. While farmer innovation and farmer investment can be driven by the necessity of increasing subsistence food production, favourable price policies are a necessary condition to provide greater incentives to accelerate the farmer-based investment and innovation processes. But this chapter also shows that favourable prices are only one necessary condition for rapid agricultural growth; they are not sufficient. In the current climate of tight or shrinking development budgets, there is a danger that budgets for the core agricultural activities of the government may be cut too much.

First of all, we have seen the important role of infrastructure investments in speeding up the process of intensification. Higher border prices in Zaire, for example, could not be transmitted to the interior of the country via the deteriorating interregional road network. And local rural roads have as much to do with the level of farmgate prices as border price. (For a discussion of the role of railway and road investments in the very successful agricultural development of Thailand, see IBRD 1983.)

Direct government investment in medium- and large-scale irrigation and drainage systems is also required. History shows that such investments become necessary long before area growth comes to a halt. In India, for example, the late nineteenth and early twentieth centuries formed a major period of investment in small-, medium-, and large-scale irrigation, but area growth continued to contribute substantially to output growth until 1965. Nevertheless, as the experience with large-scale irrigation in much of Africa shows, one can also invest in irrigation too early, when land resources are too large to make labour-intensive irrigated agriculture attractive.

In land-abundant countries, strategies based on area expansion are the lowest-cost sources of growth to even a poor agricultural population. At the opposite extreme are countries like Bangladesh, where the required resources for large-scale irrigation and drainage schemes are truly daunting and the scale of projects required exceeds what farmers, or even local governments, can undertake on their own.

Improved incentives for private research and development are also required: patent systems, other forms of protecting innovators' rights, and absence of arbitrary government interference are a must. Too many countries are excessively fearful of private seed companies, for example, or of international competition in the agricultural machinery area. None the less, historical experience and research such as Evenson's or Ruttan's shows that no country has been able to benefit from science-based technical change in the absence of an agricultural research system capable of doing both basic and applied scientific research. Even borrowing of technology has been shown to be difficult without a public-sector capacity of adaptive research. It is well worth noting that until recently most external donors have not provided support for research systems on a sufficiently long-term basis, but have detracted from the strengthening of national systems by diverting research into short-term projects. Fortunately, project trends are changing, but perhaps not sufficiently fast.

Apart from these core government activities there are areas in which the role of government is less well documented. Available evidence on extension suggests that rates of productivity return on the provision of extension activities justify the costs; and rates of productivity return on a well-managed extension system may be very high (Lau, Feder, and Slade 1984). On the other hand, governments and donors have usually overemphasized public distribution of credit since the late 1960s, probably at the expense of more important long-term investment in sources of growth. And while governments certainly have regulatory functions in areas of marketing and trade (such as the establishment of auction markets, the fostering of competition, or the stabilization of highly volatile prices), many governments have intervened excessively in the marketing and storage processes by attempting to perform functions themselves which are better performed by traders and transport enterpreneurs.

Given the amount of knowledge and research about the agricultural development process, two current debates seem rather pointless. The first is about whether it is *sufficient* to set prices right to get agricultural development going. It is not, it is only a *necessary* condition. The second debate is the one about private- versus public-sector activities. Both are required, and the knowledge base is sufficient in most circumstances to be quite clear about where private enterprise is sufficient and where it is not. Only by harnessing private initiative wherever possible and by actively pursuing the core government activities can the extraordinary rates of growth of agricultural output be achieved which are required of developing countries between now and the 2020s. Though these countries probably do not face starvation even under low agricultural growth scenarios, the goal is not avoiding starvation but raising per capita incomes and drastically improving nutrition levels.

References

Ahn, P. (1977), 'Erosion Hazard and Farming Systems in East Africa', in D. J. Greenland and R. Lal (eds.), *Soil Conservation and Management in the Humid Tropics*, John Wiley, Chichester, UK.

Allan, W. (1965), *The African Husbandman*, Oliver and Boyd, Edinburgh.

Binswanger, H. P. (1983), 'Agricultural Mechanization: A Comparative Historical Perspective', Agriculture and Rural Development Department, Research Unit, paper no. ARU 1, World Bank, Washington, DC.

—— and V. Ruttan (1978), *Induced Innovation: Technology, Institutions and Development*, Johns Hopkins University Press, Baltimore.

—— and S. V. R. Shetty (1977), 'Economic Aspects of Weed Control in Semi-arid Tropical Areas of India', Economic Program, Occasional Paper No. 13, Hyderabad.

Boserup, (1965), *The Conditions of Agricultural Growth*, Aldine, Chicago.

—— (1981), *Population and Technological Change: A Study of Long-term Trends*, University of Chicago Press, Chicago.

de Wilde, J. C. (1967), *Experiences with Agricultural Development in Tropical Africa,* vol. 2, Johns Hopkins University Press, Baltimore.

Engelhardt, (1984), 'Economic of Traditional Smallholder Irrigation Systems in Semi-arid Tropics of South India', Ph.D dissertation, Hohenheim University.

Gleave, M. B., and H. P. White, (1969), 'Population Density in Agricultural Systems in West Africa', in M. F. Thomas and G. W. Whittington (eds.), *Environment and Land-use in Africa*, Methuen, London.

Grove, A. T. (1961), 'Population and Agriculture in Northern Nigeria', in K. M. Barbour and R. M. Prothero (eds.), *Essays on African Population*, Routledge & Kegan Paul, London.

Hayami, Y., and V. Ruttan (1971), *Agricultural Development: An International Perspective*, Johns Hopkins University Press, Baltimore.

—— —— (in press), *Agricultural Development: A Global Perspective*, Johns Hopkins University Press, Baltimore.

International Bank for Reconstruction and Development (1983), 'Growth and Employment in Rural Thailand', Country Programs Department, report no. 3906-TH, World Bank, Washington DC.

Lau, L. J., G. Feder, and R. Slade (1984), 'The Impact of Agricultural Extension: A Case Study of the Training and Visit Method in Harayana, India', draft, Agriculture and Rural Development, Research Unit, World Bank, Washington, DC.

Ludwig, H. D. (1968), 'Permanent Farming on the Ukara', in H. Ruthenberg (ed.), *Smallholder Farming and Smallholder Development in Tanzania*, Weltforum Verlag, Munich.

Miracle, M. P. (1967), *Agriculture in the Congo Basin*, University of Wisconsin Press, Madison.

Morgan, W. B. (1969), 'Peasant Agriculture in Tropical Africa', in M. F. Thomas and G. W. Whittington (eds.), *Environment and Land-use in Africa*, Methuen, London.

Norman, M. J. T. (1979), *Annual Cropping Systems in the Tropics*, University of Florida Press, Gainesville.

Okigbo, B. N. (1977), 'Farming Systems and Soil Erosion in West Africa', in D. J. Greenland and R. Lal (eds.), *Soil Conservation and Management in the Humid Tropics*, John Wiley, Chichester, UK.

Pingali, P. L., and H. P. Binswanger (1984), 'Population Density and Agricultural Intensification: A Study of the Evolution of Technologies in Tropical

Agriculture', Agriculture and Rural Development Department, Research Unit, report no. ARU 22, World Bank, Washington, DC.

—— —— and Y. Bigot (1987), *Agricultural Mechanization and the Evolution of Farming Systems in Sub-Saharan Africa*, Johns Hopkins University Press, Baltimore.

Richards, A. I. (1961), *Land, Labour and Diet in Northern Rhodesia*, London.

Rounce, N. V. (1949), *The Agriculture of the Cultivation Steppe*, Longmans & Green, Cape Town.

Ruthenberg, H. P. (1971, 1980), *Farming Systems in the Tropics*, Clarendon Press, Oxford.

Ruttan, V. (1982), *Agricultural Research Policy*, University of Minnesota Press, Minneapolis.

Sanders, J. H. and V. Ruttan (1978), 'Biased Choice of Technology in Brazilian Agriculture', in H. P. Binswanger and V. W. Ruttan (eds.), *Induced Innovation: Technology, Institutions and Development*, Johns Hopkins University Press, Baltimore.

Schapera, I. (1943), *Native Land Tenure in the Bechunaland Protectorate*, Lovedale Press, Cape Town.

Toulmin, C. (1983), 'Herders and Farmers of Farmer-Herders and Herders-Farmers?, Overseas Development Institute, Pastoral Network Paper, London.

Trapnell, C. G. and J. N. Clothier (1937), *Soils Vegetation and Agricultural Systems of Northwestern Rhodesia*, Government of Northern Rhodesia, Lusaka.

Von Rotenham, D. (1971), 'Cotton Farming in Sukumaland', in H. Ruthenberg (ed.), *Farming Systems in the Tropics*, Clarendon Press, Oxford.

4 From Land Abundance to Land Scarcity

The Effects of Population Growth on Production Relations in Agrarian Economies

MARK R. ROSENZWEIG, HANS P. BINSWANGER,
JOHN McINTIRE

Department of Economics, University of Minnesota. Employment and Rural Development Division, The World Bank, Washington, DC. The World Bank, Washington, DC

There have been an abundance of analyses of the consequences of population growth for aggregate savings and investment rates and for per capita income levels and growth. However, little attention has been directed to understanding how population growth or, more specifically, changes in the relative scarcity of land and labour alter the diverse mirco-institutions which characterize the exchange (or lack thereof) of outputs, the mobilization of factors of production, and the distribution of factor endowments. Aggregative quantitative models ignore changes in institutions and, in particular, changes in the distribution of endowments. They thus do not predict the development of an economy well over a long time period. The purpose of this chapter is to account for the changes in the diverse set of production relations characterizing agrarian economies induced by long-term increases in population growth. By *production relations*, we mean the relations of people to factors of production in terms of their rights of ownership and use in production, and the corresponding relations of people among one another as factor owners and renters and as landlords, tenants, workers, employers, creditors, and debtors.

In our approach, we pay particular attention to the constraints that arise when market mechanisms for risk diffusion and intertemporal markets are incomplete and information is costly. Along with considerations of risk, risk avoidance, and information problems, we incorporate into our analysis the material features of agricultural production and the material attributes of

Prepared for presentation at the International Union for the Scientific Study of Population Seminar on Population and Rural Development, New Delhi, India, 15–18 December 1984. The views expressed in this chapter should not be attributed to the institutions to which the authors are affiliated.

production factors. Production relations and the institutions governing exchange relations are thus seen as jointly determined (1) by the aspirations of people for higher incomes and for lower risks, (2) by the physical and technological constraints encountered in a specific environment, and (3) by behavioural and technological constraints on information acquisition and transmission.

The following section of this chapter, 'Foundations', sets out the basic behavioural, material, and technological foundations of the subsequent analysis. In that section, we first consider the general consequences of risk and information costs which are unique to agriculture. Second, we introduce material and risk characteristics of agriculture and derive their consequences for the existence or non-existence of the main intertemporal markets, the market for insuring crop yields (or more generally against production risks), and the market for credit.

Third, we explore determinants of exchange and market possibilities for labour, land, and animals. These possibilities are influenced by the information problems discussed earlier in the section, by the physical attributes of the factors of production, and by the limitations on the two key intertemporal markets. We show that where intertemporal *functions* cannot be carried out through intertemporal *markets*, they must be carried out (1) via either transactions in the markets for output or for factors of production (a rationale for some tied transactions), (2) by social institutions such as the family, or (3) through accumulation of stocks of reserves and of capital.

In addition to the foundations, the study of production relations of a particular region requires specific knowlege of the agro-climatic conditions and endowments of the region. In the second part of the chapter, we discuss the consequences of population growth—of the change from land abundance to land scarcity—for production relations by first describing the 'base case' of a land-abundant and semi-arid or arid economy with simple technology, with high risk and with high transport and information costs. In that context, we discuss insurance and credit markets and relations in the markets for labour, land, and animals, including analyses (based on the 'foundations' section) of storage behaviour, livestock as an insurance device, livestock entrustment systems, extended households, and the existence of both common fields and individual fields. We then consider how production relations are transformed when population growth results in an agrarian economy characterized by land scarcity. In the land-scarce environment, again constrained by a poorly developed transport and communication network, we discuss the distribution of operational and ownership holdings, tenancy, factor rental and sales markets, the absence of animal rental markets, the importance of land as collateral, the interlinking of contracts, the dominance of 'distress' sales in the land sales market, the rental to others of land owned by small landholders, and intergenerational transactions.

Many of the individual phenomena discussed in our chapter have received explanation before. However, our goal is not to achieve a novel explanation of each agrarian relationship, but an integrated explanation of many of them, as applied to different environments. Because of this integrative approach, we do not use formalized models. This is part due to the partial nature and mutual inconsistency of the existing models of individual phenomena found in the literature.[1] Our hope is that the chapter will provide a broad framework which will assist in formalizing a model of the development of agrarian economies that is capable of explaining a rich array of phenomena and amenable to empirical application and refutation.

The Foundations

The General Consequences of Risk and Information Costs

We make the following six general assumptions:

A1. *Ubiquity of risk.* Individuals face risks from many different sources—from the production process, from the market, and from health or other factors.

A2. *Costs of information.* The acquisition and transmission of information involves costs in terms of both time and resources. Information is often acquired most cheaply as a by-product of the production and consumption activities in which one is already engaged.

A3. *Behavioural self-interest.* Individuals are primarily interested in that which they will find useful to themselves.

A4. *Consumption.* All individuals value consumption.

A5. *Effort aversion.* All individuals *dislike effort* at least to some extent, including the effort required for supervisory activities.

A6. *Risk avoidance.* Individuals will seek to avoid *risk* whenever gains and losses exceed trivial levels of income. Aversion to risk may vary among individuals, as well as for the same individual, with relative amounts of wealth.

These six assumptions have the following general consequences:

C1. *Asymmetric information.* Information has value and is costly to acquire (A2). Because individuals are self-interested (A3), they will not part with information they possess unless it is to their advantage. For example, high-quality workers want employers to have accurate information about worker quality, whereas inferior employees would prefer that worker quality be unknown to the employer (unless the employer can penalize the employees once he/she finds out). The same applies to borrowers and lenders

---

[1] For an extensive critical review of the economic literature on rural employment and contractual arrangements in low-income countries, see Binswanger and Rosenzweig (1984).

or to insurers and insured. Sellers of seeds and animals know more about their quality than do the buyers, and they may have incentives to misrepresent seed quality. Such problems of asymmetric information arise in virtually all economic transactions to some extent.

C2. *Incentive problems* (moral hazard, adverse selection, and screening effects). When information is costly (A2) and asymmetrically distributed (C1), incentive problems arise in most economic transactions. A labourer who is paid daily has no incentive to work hard unless supervised closely via direct observation of personal effort or via monitoring or inspection of individual output. Incentives to work hard may be improved by providing share contracts (piece rates at harvesting or for earth digging, or crop-sharing tenancy contracts). Because the worker receives only a share of the full marginal product of his/her effort, the worker will still not work as hard as an owner-cultivator, who receives all the rewards from his/her own effort (unless supervised or monitored in other ways and/or risking penalty in terms of loss of future repeat contracts). A person who rents a bullock pair has little incentive to feed it beyond the minimum required to elicit the work effort desired, and thus may return the animal in underfed or damaged condition. A farmer whose crop is insured against all risks relative to a 'normal' level of output will usually not apply as much care, precaution, or input as they would if the crops were uninsured.

C3. *Imperfect enforcement of property rights*. Where acquisition of information is costly (A2) and asymmetrically distributed (C1), property rights cannot be perfectly enforced. This follows from the fact that there is some positive incentive for theft that will be realized when it is easy to conceal the identity of the thief (that is, when costs of ascertaining the culprit are very high). We note here that many legal and cultural institutions are adaptations to this problem (that is, *inter alia*, they reduce costs of information or increase penalties for theft).

C4. *Desirability of a broad spectrum of insurance contracts or insurance substitutes*. This follows directly from the ubiquity of risk (A1) and risk aversion (A6). Most individuals should be willing to pay some positive amount to reduce their exposure to any one of the risks that they face. Where insurance is unavailable, they would be willing—at cost to themselves—to alter their behaviour in other ways to reduce exposure to risk (self-insurance, self-protection). Such insurance substitutes may include the holding of reserves, diversification of prospects, conservative or excessive input levels (such as the use of pesticides), investment in 'credit-worthiness', and social ties.

C5. *Collateral requirements*. At a given interest rate, collateral has three effects or functions. It increases the expected return to the lender and reduces the expected return for the borrower, it partly or fully shifts the risk of loss of the principal from the lender to the borrower, and it provides those borrowers who would have high utility under default with additional incentives to repay loans (see Binswanger and Rosenzweig 1986). The following implications follow immediately: lenders, other things equal, are more likely to lend without collateral (1) for small loans rather than for large ones, (2) to owners who have invested in land and buildings rather than to tenant farmers, and (3) to resident workers rather than migrant workers.[2]

C6. *Consequences of collateral for the existence of credit markets and for utility costs of credit.* Where suppliers of credit avoid risk by insisting on collateral, the credit market does not exist for groups of individuals who do not own assets which can be used as collateral, despite the fact that these collateral-poor borrowers may be willing to pay higher interest rates. Because the utility gain from default for the borrower increases with the interest rate charged, the expected utility for the lender may go down as the interest rate increases (Stiglitz and Weiss 1981). However, the market may also disappear for persons with no collateral from the demand side, because the expected return on the investment may be lower than the interest rate.

In environments where insurance and forward markets are incomplete, credit acts as an insurance substitute. The utility cost of identical loan terms (quantity, interest, collateral requirement) therefore differs for individuals who own different amounts of collateral: if such a loan is taken and exhausts the collateral of the collateral-poor borrower, he or she will be able to resort to additional credit (insurance) only at more adverse terms (that it, a larger 'insurance' premium than the collateral-rich borrower who still has unused collateral sufficient for another loan on the previous terms). Thus, identical loan terms do not imply identical utility costs for different borrowers (relative to the utility of labour, for example).

C7. *Forms of collateral.* Not all forms of assets are suitable for use as collateral. We define collateral as an asset which satisfies the following three conditions: appropriability, absence of collateral-specific risk, and accrual of the returns to the borrower during the loan period.

(*a*) *Appropriability.* It must be easy for the lender to appropriate the assets in case of default. This condition can be satisfied in several ways: for example, financial assets and gold can be deposited with the lender. For land, motor vehicles, animals, and human capital embodied in the borrower, ownership rights must be well defined.

(*b*) *Absence of collateral-specific risk.* Lenders must be fairly sure that the collateral cannot easily become worthless because of theft, pretended theft (in which the borrower colludes with the thief), or damage by fire, accident, or (in the case of animals) disease. Land entails little risk because of its physical characteristics. For animals, however, incentive problems make it extremely difficult to provide theft or life insurance (for health-related death causes) for animals because owners might find it

---

[2] For a more extensive discussion of collateral requirements, see Binswanger and Rosenzweig (1986).

profitable to let a sick animal die if fully insured, invest less in guarding the animals or collude with the thief.

C8. *Collateral substitutes*. People living in economies with few good collateral options or borrowers who own few assets usable as collateral either will not borrow much or will resort to inferior or more problematic forms of collateral in which the three primary conditions are not easily met. They can also, however, attempt to move to a variety of important *collateral substitutes*. One such substitute is *third-party guarantees*: a borrower who owns few assets with collateral value and whose own repayment incentives appear to be low to the lender may ask a third party who is perceived as less of a risk by the lender to guarantee repayment in case the borrower defaults. The third party may do so if they have better information about the borrower's repayment incentive/capacities, or otherwise gain from the relationship with the borrower in other than the borrowing transaction.

A second substitute is the threat of *loss of future borrowing opportunities*. For immobile populations, traditional money-lending systems transmit default information quickly to all potential lenders, and substantial lending can occur on this basis alone. The third form of collateral substitute is a *tied contract*, in which credit is given in association with another transaction. Examples occur when a trader pays in advance for a crop or provides consumer credit or when a landlord provides credit to a tenant as part of the tenancy contract; indentured servitude, debt bondage, and various forms of temporary debt sharing are tied contracts in which the labour worth of the bonded person accrues to the lender.

C9. *The valuation of assets with collateral value when insurance markets are poorly developed*. Because credit is a close substitute for desirable insurance (C4), and because assets with collateral value are often necessary for access to credit (C5), any asset with a high collateral value will be valued over and above the utility of its consumption or production stream. The full value of the asset will reflect the direct utility in production or consumption plus the utility of the insurance (via credit access) it provides. Such valuation will thus be reflected in the sales price of the asset, but not in its rental price. This is because only the owner can use an asset as collateral, whether or not it is rented; the renter cannot.

Spatial and Risk Characteristics of Agriculture and their Implications for Rural Insurance and Financial Intermediation

In addition to specifying the consequences of risk and information costs, in order to characterize production relations in agriculture it is necessary to consider the major technological features of agricultural production. One set of basic features is associated with the spatial nature of agriculture and the peculiar transport, communication, and risk configuration which results.

A7. *Spatial dispersion*. Land is an essential factor of production. Its complete *immobility* implies that co-operating factors have to be brought to it in order for production to take place, inspection of crop status for management decisions involves travel to the plot, and outputs must be transported to the point of marketing or consumption. The immobility of land leads to the characteristic geographic dispersion of agricultural production.

A8. *High costs*. Transport and travel are costly and time-intensive because of the spatial dispersion of agriculture (A7).

A9. *Seasonality*. Because of the immobility of land, yearly climatic variations of temperature, rainfall, and sunshine lead to seasonality of agriculture. These temporal variations give rise to seasonal needs for credit to bridge the gaps between receipts and expenditures. Planting and harvesting cannot be staggered to obviate this need.

A10. *Synchronic timing*. Within small agricultural regions, seasonality (A9) and the spatial dispersion of agriculture (A7) lead to high positive covariances in the timing of crop growth cycles and therefore of agricultural operations on different plots of land within the same geographic and climatic region. This gives rise to special problems of year-round rentals in labour, animal, and machinery markets.

A11. *Increased costs of information*. The spatial characteristic of agriculture and the inherent heterogeneity of factors of production increase the normal costs of information acquisition and transmission (A2). Costs of transmitting information, however, fall with the development of transport networks, printed media, and telecommunications.

A12. *Sources of agriculture-related risk*. Risk is ubiquitous (A1), but agriculture-related risks arise from four different causes or sources. Three types of risks are heavily but not exclusively associated with weather uncertainties: yield risk, market price risk, and timing uncertainties. In addition, durable factors of production may fail due to breakdown; individuals face risks of temporary or permanent absence from work because of illness or accidents.

A13. *Yield risk covariance*. Because of the spatial characteristics of land (A7), weather conditions on parcels of land within a spatial unit with similar weather conditions will have covariate yields. However, the larger the heterogeneity of land quality within the spatial unit, the less the yield covariance.

These spatial and risk characteristics have the following implications for rural insurance and financial intermediation:

C10. *Absence of crop insurance*. General crop insurance may fail to emerge for three principal reasons. *First*, given asymmetric information (C1), the costs of measuring expected yield and its probability distribution and of assessing the yield shortfall in any given year may be excessive. Fields and farmers differ significantly in expected return even in very small regions. The cost has to be charged as part of the premium. If the utility gain of the insured between the uninsured and the insured situation (that is, the 'insurance premium') falls short of the information costs of providing the insurance, no market will exist.

The *second* reason for the failure of insurance markets is the existence of incentive problems (C2)—moral hazard and adverse selection. For genuine yield insurance, the insurance contract itself reduces incentives for fertilizer application, plant or animal husbandry, and disease and pest prevention. Co-insurance clauses, deductibles, and limitation of insurance to certain risks are all means to reduce these incentive problems, but they also reduce the potential utility gain to the insured.

The yield assessment and incentives problems are both information problems which, because of costly information transmission (A2), are easier to overcome for a village resident than for a distant insurance company. Why then do some villagers not create an insurance market? The potential local insurer faces the *third* major reason for failure of insurance markets, namely high covariance of risks (A13). Because crops of all insured farmers may and do fail at the same time, the local insurer has to carry high reserves in the form of cash, gold, stocks, or short-term financial assets. In order for a village resident to write a credible insurance contract for fellow residents, the lender's reserves, at all times, have to be equal to the total insured value, and the insurance arrangement degenerates to a centralized reserve scheme. Each farmer can self-insure at the same cost (as long as the storage costs or returns from short-term financial assets are the same for all individuals). Self-insurance by holding reserves avoids all information and incentive problems, and so will usually be preferred. It is clear that a crop insurance scheme for a small region would thus not be very different from a centralized reserve system.

C11. *Problems of financial intermediation.* The problems that make crop insurance schemes costly also mitigate against the development of deposit-banking systems in rural areas. Two sets of factors imply that a local rural bank could only be an intermediary between borrowers and lenders if it had a high reserve ratio. The first set comprises seasonality (A9) and synchronic timing (A10): if depositors and lenders are both engaged in agricultural production, the depositors will want to withdraw their funds for production purposes at the beginning of the growing season (that is, at the same time that borrowers will want to borrow for production purposes as well). Deposits and repayments would similarly coincide at the harvest time.

Second, for longer-term lending, covariance of yield risk (A13) leads to covariance of default risks. In addition, covariance among yields leads to covariance among incomes of depositors and borrowers. If crops fail in one year, most depositors will find that they have low incomes and will want to withdraw their deposits. But most borrowers will find it impossible to repay loans. A money-lender who lends out of equity does indeed have a high level of 'reserves', and can reschedule loan repayments while charging interest on the outstanding balance; the expected return on such loans does not decline because of yield covariance. If instead the lender lends out of deposits, he or she may have to collect the funds via sale of the collateral, which in bad crop years may be marketable only at a discount or with other difficulties.

Costly transmission of information (A2) makes the management of a branch banking network across sufficiently large geographic zones difficult; the late emergence of branch banking for agriculture in rural areas is closely associated with the gradual improvement in information transmission.

Characteristics of Factors of Production and their Implications for Factor Rental Markets and Operational Scale

We now consider the physical attributes of three factors of production: land, animals, and people. We show how the exogenous physical attributes of the factors influence the contractural arrangements by which they can be exchanged in factor markets.

A14. *Land as a low-risk factor of production. Land* is the only completely immobile factor (A7) of the three. Once created by investments, it is difficult to damage permanently land (except where slopes lead to high erosion risk). Damage from neglect or misuse, unlike in the case of draught and breeding animals, is highly visible. Maintenance requirements occur infrequently and are often low.

A15. *Animals and animal services*. Animals, in contrast to land, are mobile and very fragile. They easily suffer permanent damage from accident, disease, maltreatment, or neglect. They correspondingly require frequent maintenance at high levels: they have to be fed or milked on a daily basis, and breeding, disease prevention, and treatment involve frequent monitoring and cleanliness. While animals are not physically divisible, the services of animals may be divisible. Production services of *draught* animals, however, always require the presence of a driver.

A16. *Workers and worker effort*. Workers' productivity is as easily impaired as that of animals and requires as much regular feeding and care, but adult workers can attend to self-maintenance. While the physical strength of a worker may be assessed by inspection, effort is not easily discerned. Therefore, incentive problems exist for effort, unless the effort is provided by the owner-operator of the farm.

Contractual Arrangement Involving Labour and Operational Scale

Labour costs have a number of components relating to the costs of supervision, maintenance, hiring, and direct payment. Given asymmetric information (C1) and dislike of effort (A5), whenever a worker is not the sole residual claimant of farm profits there is an incentive problem (C2) relating to worker effort. Minimizing or circumventing this problem requires resources (supervision costs). The ease with which the incentive problem can be reduced is thus an important cost factor associated with the use of labour. There are potentially four types of labour contracts or arrangements (we initially ignore combined land–labour contracts). The farm operator can utilize family labour, which receives a share of farm output, can hire salaried labourers on a long-term basis, hire labour for a short period (day or month) using time wages, or, whenever possible, pay workers according to the output they produce.

For short-term contracts, payments according to a worker's output—piece rates—dominate time wages: they are alike in all cost aspects except for the fact that piece rates provide superior incentives compared to time rates. However, piece rates are only feasible for operations where worker-specific output is easily measurable in both quantitative and qualitative terms. Operating a farm exclusively with workers paid by piece rates is therefore not usually feasible.

Family members, being residual claimants, have more incentives to work than hired workers; moreover, more information is available on the family workers over the long term, hiring or search costs are minimal, and family labourers share in the risk. Of course, while the incentive problem is smallest for family members, it is only incompletely solved. When marginal product cannot be precisely identified and effort cannot be discerned, shares in profits will be independent of effort. Thus, the larger the number of claimants (the greater is family size) the greater the incentive problem for each individual family worker, as shares are diluted. Nevertheless, the incentive problem will always be less than that for time-wage, hired workers. As a consequence, as the size of the labour force grows, any fixed amount of resources devoted to supervision will have smaller effects on each worker, and labour costs will rise with farm scale.[3]

Contractual Arrangements Involving Animals

Draught Animals and the Scale of Farming Operations   In the absence of other inputs in farm production, the existence of incentive problems (C2) would imply that the most common operational scale would be the nuclear family farm. Where draught animals (such as bullocks) are important, however, farm size may exceed the scale at which it can be handled efficiently by one or two persons. Because animals are fragile, requiring a high degree of maintenance (A15), and because of incentive problems (C2) for maintenance, an owner is unlikely to rent out his or her animals on a long-term basis, unless the owner is the person supervising and maintaining the

---

[3] The rise in labour costs with operational scale occurs even in the absence of labour market 'frictions' or transaction costs, and it does not require that the probabilities of finding employment or employees be less than one, conditions emphasized in the existing literature. Incentive problems (C2) are a necessary and sufficient condition for such operational diseconomies.

animals. Given the cost advantage of family labour, it is more profitable, however, for the animal owner with some level of managerial skill to cultivate land with his or her draught animals and family labour rather than to hire each factor out separately. (Depending on the precise context, the farmer can obtain land via tribal cultivation rights, tenancy, and/or ownership.) Incentive problems in both the animal and labour market thus prevent the costless circumvention of the inherent lumpiness of animal or machine inputs, leading to initially declining input costs with operational farm scale.

Other Animals   Annual rental of other types of animals is also difficult. There are two outputs to be shared: the current outputs (such as milk or weight increments), plus the health or productivity condition of the animal at the end of the year. A rental contract must cover both. It is often quite difficult to specify and measure frequent health and maintenance inputs (A15) and to monitor the productivity condition at the end. Annual rental of animals and trees is thus quite rare. For example, dairy farming often involves the renting of farms to tenants who own their own animals. Long-term rental contracts in animals are also unlikely because there are incentives for tenants to pretend loss by theft or to consume or sell the animal.

Contractual Arrangements Involving Land

Because of the immobility of land and its low maintenance frequency and because behaviour that damages the land can be easily monitored and damages assessed (A14), the impediments to the annual rental of land are less great than those associated with the annual rental of animals. Moreover, work incentives are provided if the tenant is a residual claimant of output and production risks can be shared by the tenant and owner (sharecropping). Land tenancy thus dominates the rental of all factors of production. Where land is scarce, therefore, tenancy contracts should be more prevalent than separate contracts involving the rental of animals, machines managers, and farm servants (salaried farm workers). Moreover, given the scale diseconomies associated with labour costs, where ownership holdings are highly unequal, land tenancy contracts should be common arrangement.

The Consequences of Population Growth

Production Relations in a Land-abundant Setting

To discuss the changes in production relations arising from an increase in the size of the population, we first characterize an evironment in which the population–land ratio is low, making the following specific assumptions.

   S1. *Low density.* Population density is sufficiently low that cultivable land is abundant and has no sale price.
   S2. *Indigenous use rights.* Indigenous populations have access to land use

rights at no cost or in exchange for token payments. External powers have not created property or use rights for expatriates.

S3. *Arid or semi-arid climate and crop production.* In the absence of irrigation, there is one short growing season, so seasonality is pronounced. Weather risk is high, and within small areas, yield risks are highly covariant.

S4. *Arid climate and animal husbandry.* The cheapest means of producing cattle usually involves *transhumance*, the seasonal migration of cattle among different geographic subzones. Animal husbandry has lower production risks than cropping, for two reasons. First, during minor droughts, crops may fail to produce a harvest but vegetative growth may still provide some fodder for animals. Second, in the case of local droughts, animals may be shifted to other areas unaffected by drought. The same reasons imply that covariance between animal husbandry and crop production is lower than the covariance of yields among different crops or fields in the same area. Secular droughts imply failure of both crop and animal husbandry enterprises.

S5. *Simple technology.* Technology is simple and confined to hand tools and, possibly, to draught animals. Management skills are unimportant, and technical economies of scale are limited. Gathering and hunting provide supplemental income to agriculture.

S6. *Isolation costs.* Transport and communication costs are high (the region is geographically isolated).

Market Arrangements

Labour Markets   Low population density and the associated land rights situation (S1, S2) and simplicity of technology (S5) imply two propositions:

P1. *Self-cultivation*: there is no locally resident non-cultivating labour class.

P2. There is almost no hiring or exchange of labour among resident farmers during the peak labour seasons, which in this case is the sowing and weeding season.

The reason for P1 is that because labour contracts cause incentive problems, as noted, hired labour is more costly than family labour. Easy access to land (S2) and simple technology (S5) imply that a worker's output is at least as large on his or her own plot as on an employer's plot. Therefore, given supervision costs, the employer cannot compensate a worker for the latter's foregone output on his or her own plot. Note that P1 does not preclude temporary hiring or exchange of labour; certain tasks may be easier or more pleasurable to perform in a group. The absence of hired or exchange labour during sowing time (P2) occurs because the sowing and first weeding operation are highly time-bound and synchronic in semi-arid areas. Sowing is possible only for a few days after infrequent and unreliable showers whose timing, amount, and frequency cannot be known in advance. Postponing one's own sowing to work for someone else implies a reduced expected yield because it reduces the expected length of the already short growing season. It also implies higher risks. The optimal timing of the first weeding follows from the timing of sowing: weeding delays reduce the yield and increase the weeding effort.

Widespread use of hired labour occurs primarily in two circumstances: temporary migrants can be hired from a nearby agro-climatic region with a different sowing period, or annual migrants can be hired from a poorer agro-climatic region. Labour productivity at the destination must be sufficiently high to allow employers there to compensate migrants for their foregone cultivation at home.

A direct consequence of P1 and P2 is Proposition 3:

P3. Cultivated area per working household member is largely invariant to household size or wealth.

Assumptions required for P3 are that households have access to different land qualities in roughly the same proportions, or (less likely) that land quality be uniform.

The Output Market and the Determination of Cultivated Area Geographic isolation (S6) leads to a high level of self-sufficiency in agricultural and non-agricultural commodities. Trade is limited to low-weight and low-volume items. Self-cultivation (P1) implies that, on average, everyone is self-sufficient in food. Therefore, it follows that:

P4. There is no *regular* output market in every year.

High variability of crop output does not alter P4. Because such variability is highly covariate, it does not generate *regular* opportunities for exchange between surplus and deficit households in the same year. In years of crop failure, households whose stocks run out will obviously demand output. Some wealthier households could therefore specialize in accumulating stocks for selling at high prices during years of crop failure. This option is, however, limited by several factors. Because of the absence of a labour market (P2), it is very difficult to get access to sufficient labour to accumulate stocks beyond one's own needs. Moreover, holding stocks is costly because of storage losses and limited durability of food grains. And because of weather uncertainty, any expected positive return from such purposeful stock accumulation is a risky speculative venture: stocks might have to be held through a succession of favourable years when there is no food demand.

P5. The marginal utility of effort declines sharply. Labour effort will be limited to providing for household consumption.

Households in such areas will tend to work fewer hours than in peasant societies where export opportunities or labour markets exist.

Credit Markets Collateral options are extremely limited; because land has no sales value (S1), it cannot serve as collateral. Animals are a poor form of collateral (A15), and with simple technology (S5) (given lack of Boserupian pressures for improved productivity) no important stock of machines exists to serve as collateral. The tropical climate implies limited demand for housing investment. Shifting cultivation, where practised, further limits investments in housing. The major collateral is therefore gold or other precious commodities, investments which do not directly alter output.

The limited nature of the output markets implies that marketing–credit links are not an important collateral substitute. Similarly, limitations of the labour market (implied in P1 and P2) mean that labour–credit linkages cannot provide an important collateral substitute either. The major possible collateral substitute would be the pawning of consumer durables.

P6. Limited options for collateral and collateral substitutes imply that the credit market is sharply limited from the supply side.

P7. The credit market is also limited from the demand side: this follows from simple technology (S5) and the limitations of hired labour (P1 and P2). Together, they imply little demand for working capital, which consists primarily of the subsistence requirements of household workers.

The constraints on credit supply and demand imply that the quantity of credit is small and the primary purpose of credit is for consumption.

Insurance Substitutes and Forms of Capital Accumulation

In an environment where no crop insurance exists, and where credit markets are very thin, households must find other means of self-insurance.

P8. Social institutions such as extended families and tribal groups cannot perform well as insurance substitutes for covariant risks unless they cover extensive geographic areas. They are most appropriate insurance substitutes for person- or plot-specific risks such as health risks, crop failure on a single plot, and so forth. Thus, there are incentives for households to 'extend' (via migration) geographically (Lucas and Stark 1985).

Because local social networks are poor substitutes for covariant risks, and because supply-side limitations prevent the credit market from performing an insurance substitute function, Proposition 9 follows.

P9. Capital accumulation is the major insurance substitute, and households must hold their own food stocks.

Common granaries across households would have at least a small positive benefit because yields are never perfectly covariant. However, moral hazard imposes costs that are apparently not offset by the risk-pooling benefits. Risk-pooling benefits could only be increased if lower transport costs allowed the pooling to be extended over wider areas.

As seen in the section on output markets, limited durability, storage costs, and self cultivation (P1) imply an upper bound to storage, which is related to expected consumption. Substantial wealth will therefore not be accumulated as food stocks.

Land has no value (S1). Proposition 10 therefore follows.

P10. Livestock is the major form of wealth and insurance substitute.

In addition to having a positive expected return, livestock provides the greatest risk-diversification benefits (S4), and is therefore the major insurance substitute. The mobility of livestock is an added diversification advantage, as during a local drought it can be herded to distant markets where income may not be depressed. Note that rate-of-return calculations to livestock ownership usually take account only of the mean output. By ignoring the insurance benefit, they underestimate the overall utility derived from livestock ownership.

In some semi-arid environments, the vegetation in forests and bush and grass fallows contains useful plants, providing roots, tubers, or nuts which do not deteriorate rapidly. Gathering and hunting on these common property resources therefore not only is an income source but also may be especially valuable in bad years.

P11. Common property resources may provide an additional insurance substitute.

Transhumance and Livestock Tenancy

Early in this chapter, it was shown that animal tenancy is an unattractive contract. Maintenance of the animals by someone other than the owner creates major incentive and moral hazard problems. Powerful material forces must compel the owners to choose such unattractive contracts. However, livestock tenancy is found in arid and semi-arid land-abundant areas. The necessity of transhumance (S4) provides the first material condition: a farmer cannot farm and take advantage of transhumance opportunities at the same time. The animal must therefore be cared for by another household member, a hired worker, or a 'tenant'. The farmer must invest in animals if he or she wants to accumulate capital beyond what can be kept in food stocks (P10). But why not use another household member to herd animals? This is widely done for small ruminants that are herded locally. Caring for transhumant cattle requires substantially different skills from farming. They include knowledge of watering sites, migration routes, and weather patterns over a wide region. Such skills are not easily transmitted to younger members of a sedentary farming household. Some households will find it advantageous to specialize largely or entirely in cattle herding. *Economies of scale in herding* (up to a certain herd size) imply that these herding households may find it advantageous to herd for other households if their own herds are below the optimal size.

Structuring the entrustment as a contract which involves sharing of some output rather than as a fixed fee per unit of labour or a fixed rent per animal contract provides incentives for the tenant to take good care of the animal. Moral hazard and potential theft problems imply that entrustment relationships should be of a long-term nature.

Household Size and Operational Scale

The cultivated area per worker is largely invariant to household size, *ceteris paribus* (P3). Area per worker will be governed by the following two forces: seasonality (A9) implies diminishing returns in productivity for each person's cultivation labour, as cultivated area can only be increased by planting at later and later dates; diminishing marginal utility of effort (P5) determines the limit to the cultivated area where seasons are longer and the agro-climate is less constraining.

Because the operational scale will be determined by the size of the household, it is necessary to explore the determinants of household membership. Following Meillassoux (1981), we define a *household* as a group of individuals who produce in common on at least one field and receive food out of a common food store. In the absence of economies of scale, gains from specialization, risk, and capital accumulation considerations, household size would be atomistic (that is, each person would cultivate individually). This would circumvent all incentive and moral hazard problems. We assume that there are incentives to form nuclear units. The issue addressed here is what forces lead to household extension (that is, the forming of households containing several nuclear units). We define two forms of extension: a *vertically extended household* (composed of nuclear units of succeeding generations), and a *horizontally extended household* (composed of nuclear units of siblings). A household also can be both vertically and horizontally extended.

We hypothesize that extended households (particularly for vertically extended households) provide an insurance substitute independent of that based on occupational (Cain 1981) or geographical diversification. There are two types of insurance benefits. First, contributions to a common granary can insure members against individual-specific risks: loss of output arising from specific crop failure or from a loss of cultivation labour due to illness or accident. These risks are not covariant, and risk-pooling benefits are substantial. Both horizontally and vertically extended households can assume this function. A second insurance function arises where older individuals have accumulated assets to which they have exclusive claim. They provide insurance substitutes to younger household members against covariant risks such as area-wide crop failure. This is Meillassoux's explanation (1981) of vertical extension. Note that each sibling can insure just as well against covariant risks if he or she forms a separate household with his/her wealth share.

The requirement of producing on a common field arises as follows: insurance via common stocks involves moral hazard. The manager of the grain store must have enough information to determine that the plot- or indivi-dual-specific misfortune occurred despite the best efforts of the insured. This is particularly difficult in the case of plot-specific crop failure, as the insurer cannot easily observe the effort spent by the insured on the individual plot. A requirement to produce for the common granary on a common field circumvents these information and moral hazard problems. Equal effort is enforced by mutual supervision on the jointly cultivated fields.

Individual ownership of wealth and control over common food stocks provides older individuals with powerful means of extracting labour services from younger family members, a central point in Meillassoux's explanation of production relations in semi-arid, land-abundant areas. If the insurance benefit provided by the household head is not counted, the expected con-tribution of children to output and household wealth will exceed the pay-ments they receive (that is, their consumption out of the common store). The household head provides insurance in exchange for cheap labour, which might enable the household to accumulate faster than it otherwise would. An added incentive for vertical extension arises from the fact that older people have a greater knowledge of variability in production conditions (Rosen-zweig and Wolpin 1985). This would be particularly relevant in the highly variable arid and semi-arid climates and would also justify joint production.

We note that none of the preceding arguments imply that households must be formed exclusively by family members or close relatives. Household members also may choose not to cultivate *all* their plots in common. The insurance arguments only require a portion of output to be produced jointly which is intended for storage in the common granary. These explanations, moreover, do not account for the co-residence of joint producers or their consumption from a common kitchen. However, they do suggest the forces limiting the size of households: the limiting household size is reached when marginal insurance benefits from risk pooling fall short of the work-effort incentive-dilution effects.[4]

Effects of Population Growth: Production Relations in Land-scarce Settings

In a closed region, population growth changes many characteristics of the low-population-density crops. In Boserup's (1965) analysis, population

---

[4] An alternative explanation of the extended family has been developed by Kotlikoff and Spivak (1981). The age at which death occurs is uncertain. Where capital and insurance markets are well developed, individuals can invest their capital in an annuity which provides them with a fixed income for the rest of their life, no matter how long it is. Where annuities markets do not exist, they can enter into a contract with younger individuals to bequeath their wealth to them in return for their support during that period of their life in which they are not able to work. Where absence of annuities markets *may* reinforce a vertically extended household, the model does not provide an explanation for joint cultivation and consumption out of the common granary as do the insurance and experien-tial hypotheses. Nor does it require that the transactions be between blood-relatives.

growth has eight principal effects: (1) it reduces the fallow period; (2) it increases investment in land; (3) it encourages the shift from hand-hoe cultivation to animal traction; (4) it encourages soil fertility maintenance via manuring; (5) it reduces the average cost of infrastructure; (6) it permits more specialization in production activities; (7) it reduces a change from general to specific land rights; and (8) it reduces the per capita availability of common property resources (forest, bush, and/or grass fallows; communal pastures).

The first four effects listed result from the necessity to raise land productivity and to offset the increase in labour requirements associated with more intensive cultivation. The fifth and sixth effects are due to economies of scale of providing infrastructure in more densely settled areas, and the concomitant increase in intraregional trade.

The seventh effect, transition to specific land rights, while subject to enormous regional variations, has a common tendency. With general land rights, cultivators typically own only the right to cultivate in a particular region. A lineage head assigns the right to use a specific plot and to do so as long as they actually cultivate it; when the current cultivator departs—usually to leave the plot fallow—the use right to the plot reverts to the lineage. With the development of specific land rights, the cultivator can begin to assert certain rights in plots, starting with the right to resume cultivation of the specific plot after a period of fallow. At a later stage, the cultivator asserts, and will receive, the right to assign the plot to an heir or to a tenant; the use right in the plot does not revert to the lineage any more. With increasing population density, the rights assignable by the individual cultivator become more extensive and eventually include the right to refuse stubble grazing and, most importantly, complete alienability—the cul-

Table 4.1   Comparisons of production relations: low and high levels of population density

Characteristic	Land-abundant economy	Land-scarce economy
Purchased inputs	Few	Many
Specific land rights	None	Well developed
Common property resources	Abundant	Scarce
Collateral value of land	Absent	Important
Land sales	None	Rare
Demand for credit	Low	High
Money-lending	Low	Well developed
Landlessness	None	Exists
Local labour markets	Confined to off-peak periods	Well developed
Land tenancy	Present	Pervasive
Livestock tenancy	Present	Absent
Horizontal extension of households	Pervasive	Rare
Vertical extension of households	Pervasive	Pervasive where technology stationary

tivator can lease and sell plots to individuals from outside the lineage. This transition to secure specific land rights provides incentives to undertake investments into specific plots, investments that are required for the intensification of production and preservation of soil fertility.

Apart from these general effects, population growth that leads to a higher population density has additional consequences for risk and for production relations: for the levels of credit, for rental and sales markets for factors, for asset distributions, and for family structure. The additional effects may vary according to the degree of technical change and land investment induced by population growth itself. Table 4.1 summarizes and compares production relations across low and high population density regimes.

Population Growth and Risk

Population growth has little effect on person-specific or covariate risk, but it has important effects on the means to diffuse them. Population growth mitigates person-specific risk in two aspects. One, it leads to markets for specialized crops (such as fruits and vegetables) and allows greater diversification into them. Two, it reduces the average cost of infrastructure, thus allowing further market expansion for some goods and further diversification.

Population growth diffuses covariant risk by reducing the average cost of infrastructure. Because it costs less to trade with other nearby areas in bad years, there will be intraregional gain from risk reduction even if the region is not open to external trade. This is not the same thing as the second form of reduction in person-specific risk, because it occurs independently of changes in individual behaviour. The reduction in covariant risk, imputed to a reduction in costs of access to less covariant activities, would occur even if the area produced everything it did in the same proportion before the cost reduction. The reduction in person-specific risk, while it has the same indirect cause, depends on the individual response to new intraregional trade opportunities created by population growth.

Population Growth, Land Sales, and Landlessness

With induced innovation and investment, the demand for both working and investment capital rises, as does the demand for trading credit to accommodate increased intraregional specialization. A major consequence of increased population size, however, is that land acquires scarcity value and thus, due to its inherent characteristics, value as collateral. The availability of a new collateral asset sharply increases the supply of credit but has consequences as well for the liquidity of the land sales market, for the distribution of land-holdings, and for the existence of a landless population.

Two features of agriculture that we have stressed—the high positive spatial covariation of income and the value of land as collateral where land is scarce—imply that in periods of normal weather conditions there would be few land sales. On the supply side, sellers of land would only be better off if they could earn a higher return from the proceeds of the land sale than from self-cultivation or land rental. The supply of land for sale will thus be small where non-agricultural investment opportunities for rural residents are limited and national credit markets are undeveloped. In many instances, the only alternative investment would be money-lending, which would not allow the potential seller to escape the risks associated with agriculture and would entail the supervision of debtors rather than workers or tenants.

On the demand side, the number of bidders for land will be less con-strained by the level of self-generated savings because the mortgaging of land would be unprofitable. Consider a landless labourer who wishes to purchase land by obtaining a loan and using the land purchased with the loan as col-lateral. Because (unpledged) land has collateral value (lowers net credit costs), the equilibrium price of land will always exceed the present dis-counted value of the income stream produced from the land, for *given* credit costs. The owner of mortgaged land, however, cannot use the land as col-lateral for working capital, does not reap the production credit advantage, and thus will be unable to repay the loan for the land purchaser out of the increased income stream. Only land not already pledged as collateral yields a flow of income whose present value equals the land price. Land sales thus are likely to be primarily financed out of self-generated funds, from savings, so that the purchased land can be used as collateral for working capital. The necessity of purchasing land out of savings might thus tend to make any given distribution of land-holdings more unequal, despite the greater utility value of land (insurance value, collateral, lower labour costs) to smaller owners.

The covariation in yields suggests that in particularly good crop years, when savings are high, there would be few sellers of land and many potential buyers at the 'normal' price; good years are thus not good times for land pur-chases. In bad years, there would be little savings to finance land purchases; indeed, in particularly bad periods of weather—consecutive harvest failures, for example—money-lenders would be the only individuals with assets (their debt claims). Money-lenders would prefer to take over rather than to sell the land-holdings offered as collateral by defaulters, because the price of land would be lower than average in bad years. Thus, in bad crop years, *distress sales* (or transfers of land at values set at the time of the loan) would be mainly to money-lenders and would be the most prevalent transaction.

The emergence of the credit market based on the collateral value of land thus leads to the emergence of a class of money-lenders and a class of non-owners of land. Money-lenders increase because increased collateral options make money-lending more profitable in the short run. Moreover, as dis-cussed, money-lending becomes a way to acquire land pledged as collateral. A class of non-owners arises because land pledged as collateral is sur-rendered on default. A positive frequency of default arises because for each individual it is not optimal to pursue a borrowing strategy that implies zero probability of involuntary default. Once land is lost, it cannot be bought again with borrowed funds. All the preceding tendencies are stronger with induced investment than without.

Operational Scale and the Labour Market in a Land-scarce Environment

Given the availability of labour for hire in a land-scarce setting, the cost-minimizing operational scale will no longer be determined by the amount of cultivation that can be performed by family members. When there is marked seasonality of farm operations and labour demand (A9), given the already committed costs associated with the consumption by family members during the year, it is optimal for hired labour to be used in peak periods. That is, the optimal permanent labour force will be less than the peak labour demand (the peak-period marginal product of family labour should be greater than the peak-period market wage prior to hiring). Conversely, given hiring costs, two factors make reliance on hired labour alone risky: synchronic timing of operations across farms (A10) and the importance of timeliness of operations (A9). Therefore, the optimal permanent labour force will be greater than the minimal or off-peak quantity of labour demand. In the absence of off-farm employment by family members, the slack-period marginal product of family labourers would therefore be less than the slack-period wage. Thus, where there are no slack-periods, off-peak jobs there will be seasonal overpopulation and underemployment. Thus, seasonal underemployment, the rise of labour cost with operational holdings, and the relationship between operational scale and the size of the family labour force are joint consequences of the behavioural and technological characteristics of labour arrangements.

It is important to note that the rise in labour costs with operational farm scale occurs even in the absence of frictions (hiring costs) in the labour market, and even if probabilities of finding short-term employment or employees were equal to one. *Incentive problems (C2) are necessary and sufficient conditions for the rise in labour costs with operational holdings.*[5]

Land Tenancy, Operational Scale, and the Ownership Distribution of Land

In a dense-population, land-scarce environment, the value of land as collateral implies that owners of large amounts of land will be able to acquire credit more cheaply than owners of smaller amounts. Does this production

---

[5] There are exceptions to our claim that farms operated principally by salaried managers using hired farm workers are inefficient compared to farms managed and worked by family members with claims on profits. These exceptions (large-scale, hired-worker operations, i.e. plantations) arise chiefly from two sets of technological conditions that are associated with the crop grown. Both of these conditions make it very difficult to use tenancy (i.e. to take advantage of lower-cost family labour). Plantation crops are either (1) long-term crops with high maintenance intensity, such as trees and/or (2) crops where there are important scale economies in processing *combined* with co-ordination problems between harvesting and processing (sugar-cane). See Binswanger and Rosen-zweig (1986).

cost advantage of large ownership holdings imply that the distribution of land ownership will importantly affect operational scale?

Our analysis implies that a landless person with a family who owns animals and/or machines and possesses some managerial skill will find it more profitable to rent land than to hire out his or her endowments separately. Similarly, a large landowner will find it more profitable to rent out land than to manage a large operation, because of the scale diseconomies arising from the use of hired workers. By renting out parcels of land to individuals using primarily lower-cost family labour, landlords with larger land-holdings gain; by renting land, households with little or no land gain. There is thus a potential market for the rental of land, which would tend to make operational holdings more equal than ownership holdings and to move the former closer to optimal operational size. Moreover, the lower credit costs faced by owners of large land-holdings can often be passed along to tenants. *To the extent that this can be done, the pattern of ownership holdings has little effect on overall production costs.*

The existence of a land rental contract between a landowner and a tenant is likely to entail the provision of credit by the owner to the tenant. This is because of economies of scale in monitoring; because the owner is more likely to have superior information on the tenant as compared to other individuals; and because the cheaper credit can be embodied and even disguised in owner-provided inputs which help ensure their proper use. The linkage of contracts (that is, transactions in more than one input between tenant and landlord) is thus likely to be common. It represents a means of achieving lower production cost where the distribution of ownership holdings is highly skewed and there is a more homogeneous optimal operational scale.

To the extent to which tenancy arrangements are feasible, there is an indefiniteness to the distribution of ownership holdings. While transaction and supervision costs rise with the number of tenancy agreements involving optimal-sized plots, and thus with the size of the ownership holdings, the rise in costs with ownership holdings is likely to be small. For very large ownership holdings, specialization in rent collection is likely, and use of family members as supervisors of tenants reduces principal-agent problems. At the low end of the distribution, owners of holdings smaller than the optimal scale who also own animals will rent land from larger (lower-cost credit) owners. Owners of small holdings without animals or management skills, however, will rent out land to larger landowners, because such owners will want to preserve some collateral as insurance substitute and cannot therefore pass on credit to a pure tenant. This requirement of capital ownership implies that the non-operating, landless worker group will be larger with induced innovation and investment than without.

In sum, we would expect that in land-scarce areas with poorly developed capital markets, land sales would be few and limited mainly to distress sales.

There would be a tendency for land to accumulate in the hands of persons with greater production endowments—managerial skills, family labour, bullocks—but this also means that land would tend to accumulate faster among persons with already high land ownership. Thus while both the land rental and sales markets promote the efficient allocation of factor endowments, land sales are likely to exacerbate land-holding inequality, necessitating greater use of tenancy contracts. Land sales and land rentals evolve as complementary transactions.

Livestock Ownership and Tenancy in the Land-scarce Environment

As land ownership becomes a major form of investment and of insurance substitution, the role of animals as insurance substitutes declines. As renting out land involves fewer incentive problems than renting out animals, livestock tenancy will decline in those agro-climates where it was important. In addition, higher population density leads to increasing conflict over common properties and reduces their extent. Opportunities for transhumance decline, and the necessity for a tenancy contract in livestock is lessened. As is well known, increased population density usually a shift towards storing and producing fodder.

Household Structure, Population Density, and the Land Market

Improvements in land, credit, and output markets associated with population growth will lead to a decline in the prevalence of horizontally extended families, as such markets provide superior opportunities to diffuse person-specific and even covariant risks. However, in the absence of rapid technological change, if plots of land are differentiated and weather variability is high, experience specific to plots of land has a high return. Rosenzweig and Wolpin (1985) demonstrate that these returns to land-specific experience make it profitable for generations of kin to work together on the same plots of land and thus reinforce tendencies toward the vertical extension of families. The specificity of experience also makes sales of land to non-kin inferior to bequests of land to offspring. Thus, sales of land are rare and tenancy contracts will often involve the same individuals from year to year.

Conclusion

The survival and growth of societies depends on the generation of a set of relationships in production governing exchange and ownership that allow individuals or households to achieve their individual or collective goals effectively in the presence of risk. Production relations therefore should be expected to adapt, however imperfectly, to the current and intertemporal problems that arise as a result of changes in material conditions. In this chapter we have attempted to delineate the principal behavioural and technological factors which act as determinants of production relations in agriculture in order to assess how such relations are transformed as a consequence of the shift from land abundance to land scarcity associated with population growth. Our analysis departed from many of those in the existing literature in its application of the general risk and information problems (which have been the focus of recent work in micro-economic theory) to the unique technological characteristics of agriculture in order to provide an internally consistent explanation of many well-documented institutional features of land-abundant and land-scarce economies with poorly developed transport and communication networks.

Apart from the material and behavioural elements considered in our analysis, we did not consider several other possible determinants and/or goals and decisions. We also excluded from consideration the endogenous determination of state actions such as conquest and coercion or legal codes. Indeed, we took certain actions of the state, such as the protection of property rights in land and the absence of slavery, as given. A complete theory of production relations will attempt to analyse such major state interventions as *partly* endogenous responses to material conditions. It would do so recognizing that there are other exogenous influences which transcend the material conditions of a specific historical rural setting.

References

Binswanger, H. P., and M. R. Rosenzweig (1984), 'Contractual Arrangements, Employment and Wages in Rural Labour Markets: A Critical Review', in H. P. Binswanger and M. R. Rosenzweig (eds.), *Contractual Arrangements, Employment and Wages in Rural Labour Markets in Asia*, Yale University Press, New Haven, Conn.

—— (1986), 'Behavioural and Material Determinants of Production Relations in Agriculture', *Journal of Development Studies*, April, 503–39.

Boserup, E. (1965), *The Conditions of Agricultural Growth*, Aldine, Chicago.

Cain, M. (1981), 'Risk and Insurance: Perspectives on Fertility and Agrarian Change in India and Bangladesh', *Population and Development Review*, 7(September), 435–74.

Kotlikoff, L. J., and A. Spivak (1981), 'The Family as an Incomplete Annuities Market', *Journal of Political Economy*, 89(April), 372–91.

Lucas, R. E. B., and O. Stark (1985), 'Motivations to Remit: Evidence from Botswana', *Journal of Political Economy*, 93(October), 901–18.

Meillassoux, C. (1981), *Maidens, Meal and Money: Capitalism and the Domestic Community*, Cambridge University Press, Cambridge.

Rosenzweig, M. R., and K. Wolpin, (1985), 'Specific Experience, Household Structure and Intergenerational Transfers: Farm Family Land and Labour Arrangements in Developing Countries', *Quarterly Journal of Economics*, supplement, 961–88.

Stiglitz, J. and A. Weiss (1981), 'Credit Rationing in Markets with Imperfect Information', *American Economic Review*, 71(June).

5 Population Growth and Agrarian Outcomes

MEAD CAIN AND GEOFFREY McNICOLL
Centre for Policy Studies, Population Council, New York

The effects of population growth on agricultural development are typically portrayed along lines of Malthusian pessimism or Boserupian optimism, with occasional syntheses that bravely straddle this gulf. We argue in the chapter that such models, attempting to isolate a simple demographic–economic system in which population-induced changes in technology and institutions can be traced out, misstate the effects of population growth in most circumstances of interest in the contemporary Third World and ignore the major determinants of agrarian outcomes. The undoubted importance of population growth in such historical transitions as the establishment of permanent field cultivation or the decline of the manorial system does not imply the existence of large routine demographic effects on the structure and thence the performance of the agricultural economy. Agrarian outcomes defined in terms of the rate of growth of per capita product are to a considerable degree tied to forms of institutional arrangements in two areas neglected in most discussions of population and agrarian change: the family system and local-level community and administrative organization. These institutional forms are often relatively immune to population growth, and indeed they may have extraordinary persistence in the face of economic and demographic change. Their origins in most cases far predate the modern era of population growth, deriving from fundamental needs for societal continuity and security. Under certain configurations of such institutions, population growth induces productivity improvements that accommodate or feedback effects that tend to restrain that growth. Under other configurations, perhaps once equally 'satisfactory' in meeting societal needs, no such responses are generated; instead, the system rachets itself to higher and higher levels of population density without commensurate product growth in an all-too-familiar process of immiserization.

This is not to argue that agrarian outcomes are fully determined, because labour, investment, and technology in the agricultural economy are in part governed by factors exogenous to it, and institutional change bearing on this economy can come about in a wide variety of ways (including by deliberate

Prepared for the Seminar on Population, Food and Rural Development, New Delhi, 15–18 December 1984.

or fortuitous policy measures) to offer escape from such an economic and demographic impasse. Nor do we claim that, within a given institutional configuration, population growth effects on the *distribution* of wealth or income are minor (although that may in fact typically be so). But the case set out here does suggest a need to rethink the role of population in agrarian change. Plausibly the most significant effect of rapid population growth is not any simple contribution to or detraction from productivity but rather that of altering the social and political costs of purposive institutional change, sometimes perhaps easing a necessary transition but more commonly raising those costs, even to virtually prohibitive levels.

Mechanisms of Population-induced Agrarian Change

Malthus has frequently been castigated for failing to recognize the first stirrings of technological change that was to transform the agrarian economies of his time. Yet his predictive failure, such as it was, derived not from any inability to imagine positive responses to the potential threats to livelihood of unrestrained population growth, but rather from a dim view of the likelihood that most people would as a general rule avail themselves of those responses. (J.S. Mill, with the same theoretical bent as Malthus but a rosier view of his fellow man, saw a future of steady economic progress.) It takes little complicating of the Malthusian system, however, and no greater faith in mankind, to make population growth itself an all but irresistible force of economic change.

Historical research and contemporary observation leave little doubt of the determinative role of population growth in the adoption of sedentary cultivation. With increasing population density the ecological stability of swidden (slash-and-burn) agriculture is lost, as fallow periods become insufficient to permit regeneration of forest cover. Natural soil fertility is progressively exhausted, the land subject to erosion and laterization and to the invasion of savannah grasses. The swidden cycle cannot continue, and sedentary cultivation (the techniques of which may have long been known) is adopted. Gourou (1965) and Bartlett (1955–61) document numerous instances of this transition. In some societies a complex body of customary law developed to assign and keep track of the slowly solidifying rights of cultivators to permanent occupancy of former swidden land.

On evidence such as this, Boserup (1965) drew her broad conclusions about density-induced productivity improvements in agriculture. In essentials, sustained population growth produces a situation of labour abundance and land scarcity with consequent effects on relative factor prices, and induces efforts to establish contractual arrangements that permit more exclusive use of land. Some form of enclosure takes place; incentives for investment rise sharply; and land yields increase.

The same argument can be applied to another major agrarian transition:

from the European manorial system of lord and villein to landlord and tenant or to freeholding peasant proprietors. This is done in a well-known study by North and Thomas (1971, 1973). Closure of the land frontier in Western Europe occurred in the late Middle Ages. Continued population growth thereafter, North and Thomas argue, led to a restriction in common property use of land and enabled the manorial lord to demand greater labour time or payment in kind for his protective and judiciary services. Property rights in labour were attenuated, property rights in land strengthened. Such changes, of course, were slowed by the inertia of custom. 'We would not expect a once-and-for-all jump from common property to fee-simple private property: in terms of political and military strife the costs of abrogating the conflicting customs of the manor would have been prohibitive. Rather we would expect successive steps to reduce freedom of entry and to increase the degree of exclusiveness in land use' (North and Thomas 1973, 23). (A host of second-order considerations that disguise these gross effects and introduce regional variations need not concern us here.)

The classic manor in the North and Thomas model is seen as a contractual arrangement suited to a situation where factor and product markets barely exist and where enforcement costs (dealing especially with problems of shirking accurate determination of output) militate against sharecropping, fixed wage payments in kind, or fixed rents in kind (North and Thomas 1973, 31–2). The gradual development of product markets and financial intermediaries from the thirteenth century onwards spread a generalized knowledge of prices and offered substantial market-scale economies to be reaped, lowering the cost of market-based transactions. Market development in turn can be traced at least in part to the effect of population growth 'accenting the differences in factor endowments between regions' and hence enlarging the gains from trade (ibid., 50–7).

As these two examples might suggest, large shifts in agrarian production relations and in their associated forms of contract are rare events. For the most part, declines in wage–rent ratios result in familiar shifts along production possibility frontiers, with substitution of labour for capital, and perhaps in capital-saving innovation. (Distortions in factor prices and the 'compulsive sequences' that in part govern technological change—see Rosenberg (1976, 111)—may of course interfere with such responses.) However, there are also effects of population growth with more routine potential impact on agrarian institutions. Tracing out these effects is becoming a popular topic of research (see, for example, Binswanger and McIntire 1984, Hayami and Kikuchi 1982). Practitioners follow North and Thomas in focusing on demographic influences on transaction costs in factor and product markets; unlike North and Thomas, however, they are concerned to explain the minutiae of institutional adjustment rather than the transformation of whole systems.

Hayami and Kikuchi's treatment of population growth in high-density wet-rice agrarian systems in Asia illustrates the genre. The initial condition here is one of pervasive sharecropping, seen as representing 'a saddlepoint between the tenant's strong risk aversion and the landlord's calculation of transaction cost' (1982, 34). Despite improvements in technology and infrastructure, labour demand often cannot keep abreast of labour supply as rapid population growth continues, resulting in declining returns to labour per unit of land and increasing returns to land. This process promotes further concentration of land, leading toward an agrarian system of large commercial farmers and a landless proletariat. Acting against this polarization, however, are the high transaction costs of wage–labour contracts in comparison with either family labour or traditional patron–client ties where contracts are informal and embedded in continuing multi-faceted relationships. One outcome observed by Hayami and Kikuchi (1982, 53 n) in several South-east Asian settings is the institutional adjustments within patron-client systems that tend to produce allocative outcomes that mirror neo-classical marginal returns while preserving traditional forms. Whether the resulting arrangements are robust enough to resist pressures of polarization remains very much in question.

Binswanger and McIntire (1984) adopt a more stylized approach, focusing not on particular historical cases but rather on the logical consequences of population growth for production relations in agriculture, given certain behavioural postulates and the existence of risk and information costs. The 'base case' they consider is that of a semi-arid or arid economy with simple technology where land is virtually a free good. The implications of population growth for the evolution of factor and output markets are then traced out. For example, with increasing population density, land eventually acquires scarcity value and, thus, value as collateral. This emergence of a superior collateral asset dramatically increases the supply of credit; the development of credit markets, in turn, has additional consequences, including the creation of money-lenders and a class of landless people.

While differing in ambition, these various attempts to model the influence of population growth on agrarian institutional change clearly share many features. They interpret institutional arrangements as locally efficient ways of coping with risk and transaction costs under given production conditions. Population growth alters those conditions through a variety of means (changed factor proportions, scale effects, savings behaviour, induced technical change, and so on), in turn yielding a different array of risks and transaction costs and making an alternative institutional configuration a more efficient arrangement. It does not follow that the system will abruptly shift to the new configuration, because the change itself is likely to entail political or economic costs; but the long-run likelihood is that shift will occur.

In actual current or recent instances of agrarian change, disentangling population-induced effects from other changes in the system usually involves a large measure of speculation. Typically, the situation is one where new agricultural techniques are being introduced and there is growing use of modern inputs—in both cases, not strictly demand-generated phenomena. Government extension programmes expand over an ever-greater range of activities. Education is spreading; knowledge of the world penetrates even remote areas through radio and television. Each of these changes may have economic effects. Moreover, the concept of population in agriculture, as defined simply by occupation or industry division, loses the limited value it once had as rural families increasingly come to resemble miniature highly diversified conglomerates, many of them with a foothold in the urban sector. Within this welter of social change, there are necessarily massive *ceteris paribus* assumptions in seeking to identify and explain intrinsic system responses to population growth.

Institutional Determinants of Agrarian Outcomes

There is, however, a more fundamental limitation to the kinds of studies we have been describing, if they are seen as more than *ad hoc* explanations of observed patterns of change in particular situations. While an interest in agrarian outcomes naturally requires a focus on agrarian institutions, to restrict the field of vision simply to those institutions is to omit elements of the institutional structure that may critically influence those outcomes. In particular, we shall argue, both the family system and the forms of local-level community and administrative organization are major determinants of agrarian success or failure. Certain forms of such institutions go far to guarantee positive results—that is, the long-run expansion of per capita product and population growth brought down to moderate or low levels. Other configurations all but ensure the opposite outcome: stagnant or falling productivity and continued rapid population growth. In the discussion following, we refine and defend this hypothesis. If correct, most of the distinctions that 'new institutional economists' such as Hayami and Kikuchi try to explain (in the details of the tenure system and labour contracts, in the timing of emergence of financial intermediaries, and so on) are second-order issues. They concern the fine structure of agrarian sytems whose gross productivity performance is in large measure determined elsewhere.

Of course, family and community systems, and to some extent local administrative structures, are societal inheritances, and they usually and properly are seen as fixtures rather than matters for policy deliberation. To recommend a correct choice in such domains is like medical advice that recommends careful choice of one's grandparents. Abstracting from such difficulties does not remove them, however, and in the present case there are some policy avenues available, which are mentioned briefly at the end of this chapter.

Family Systems

For the purposes of this argument, there are three broad kinds of distinction to be made among family systems: how they deal with intergenerational property transfer; their control over establishment of new households; and their marital fertility responsiveness to the changing economics of children.

First, family systems can be distinguished by the success or failure with which they preserve the integrity of agricultural holdings from one generation to the next, through rules of property devolution, in the face of varying rates of population growth. Systems of primogeniture, such as prevailed in much of pre-industrial north-west Europe and Japan, are relatively successful in this respect, in contrast to the practice of partible inheritance which is typical of South Asia—and is becoming typical of Africa with the emergence or strengthening of individual land ownership rights.

Family systems can be further distinguished according to whether and by what means they provide mechanisms for adjustment of population growth in response to varying economic conditions. In pre-industrial north-west Europe, the rules of family formation permitted adjustment through the timing of marriage. According to these rules, marriage marked the establishment of an independent household, for which a prior accumulation of capital was necessary. Many young people acquired this capital through employment in domestic service (Kussmaul 1981). Times of economic hardship saw the period of service extended and marriage delayed. Similar (probably more stringent) controls of new household formation existed in pre-industrial Japan, associated with the high value accorded preservation of the family land-holding and descent line and perhaps also with collective village responsibility for land tax. In this way, Japan experienced near constancy of rural population over generations. (See Fukutake 1972, Hanley and Yamamura 1972, T.C. Smith 1977.) No comparable mechanism of response and adjustment exists for the joint family formation systems once prevalent elsewhere in Europe, and still so in most of contemporary Asia and the Middle East. There, the timing of marriage is not responsive to variations in economic conditions, because marriage is not coincident with the formation of an independent household: a newly married couple typically is sheltered for a period as a member of an elder's household (Hajnal 1982). Nor are there economic constraints on early marriage or household establishment in traditional, lineage-dominant family systems of Africa—nor appreciable signs of such constraints appearing as these systems are increasingly stressed.

The family systems of pre-industrial north-west Europe and Japan, however, were never exposed to the kind of rapid population growth that has been experienced by less developed countries since the 1950s, and it is certain that, if they had been exposed, the 'marriage value' would not have provided an adequate adjustment. In the face of modern-era rates of population growth, adjustment must entail some reduction in marital fertility. This brings us to a third criterion for distinguishing family systems: their implications for the economic value of children to parents and, thence, for the extent to which secular mortality decline alone will produce a private incentive for reductions in marital fertility. The discussion here is necessarily somewhat speculative.

Particular family systems appear to have distinct structures of incentive with potential bearing on marital fertility. This is most evident with respect to the security and welfare of elderly parents. In societies of north-west Europe, from a very early period in history, there seems to have been little connection between the number of surviving children and the welfare of parents in old age (Gaunt 1983, Hajnal 1982, R.M. Smith 1984). For those with property, an explicit retirement contract was the typical means of securing food and other necessities in old age until death. The existence of such contracts suggests the limits of unenforced filial obligation; it also points to the fact that the contracted partners of the elderly need not have been children (although children may have been preferred) or even kin. For those without property, the prospect of ageing appears to have been predictably grim. The evidence for England, for example, suggests that for those without property a pattern of life-cycle poverty produced a regular disparity in economic status between an elderly parent and his or her children (R.M. Smith 1984). For this class, the fate of the elderly rested with publicly provided relief for the poor. Under these circumstances, one may conjecture that concerns about economic security in later life did not figure prominently in reproductive decisions or behaviour. In the event of secular mortality decline, it is similarly unlikely that security concerns would affect the prospect of a compensating adjustment in marital fertility.

In the joint family system, however, there is a strong connection between children and the economic well-being of aged parents. In fact, co-resident children are the system's solution to the dependency problems of the elderly for both propertied and propertyless alike. A secular mortality decline would be experienced rather differently under the joint system. Depending on how security needs are defined, the proliferation of surviving children following the mortality decline may well be perceived as a windfall gain: security in old age, which before the decline was as uncertain as child survival, seems reasonably assured after the decline. With respect to adjustment in marital fertility, therefore, the path to demographic equilibrium may well be blocked as long as the family remains the dominant welfare institution or until financial markets develop to an extent that annuities become widely available (or a public agency intervenes) (Cain 1983).

Among societies that have a joint family system, one further distinction must be made relating to how security needs are defined. In many such societies, it is not children, undifferentiated by sex, who are looked to by parents for security in old age but rather sons. In other societies, children of either sex may satisfy security requirements. For any given mortality decline (and other things being equal), one can expect a larger adjustment in fertility in societies of the latter type than in societies where the preference for sons is strong. This follows from the simple observation that fertility will be higher if parents set a goal of two sons (for example) than if their goal is two children of either sex.

Marital fertility is also seemingly insulated from economic change in African family systems. In the dominant patriarchal variants of such systems, children become members of the husband's lineage, but their subsistence is largely the responsibility of their mother. The mother, however, as in the joint family case, later may expect to have to rely on her children for support in her widowhood and old age. Whether the decay of lineage influence will lead to a family-level incentive structure that promotes a marital fertility response to the effects of population growth is a critical question for Africa's future.

Under conditions of land scarcity, the joint family system thus appears to be less conducive to successful agrarian outcomes than were the systems that evolved in pre-industrial north-west Europe and Japan, particularly in the face of rapid population growth. Partible inheritance promotes a splintering of agricultural holdings; the system lacks a mechanism comparable to the European 'marriage value' or its Japanese equivalent; and the family as the locus of welfare functions creates a structure of incentives that inhibits the responsiveness of marital fertility to secular mortality decline. *Mutatis mutandis*, much the same can be said about African family systems.

Community and Local Administration

The second component of a society's institutional structure that we would argue can have a strong effect on agrarian outcomes is the local-level social organization above the family—in particular, the characteristics of its community and administrative structure. The stereotype of traditional rural society posits solidary hamlets or 'natural villages' as the next important social unit after the family. Though the *gram raj* or village republic is no longer taken seriously as a depiction of historical reality, in most rural societies local supra-family groupings did exert considerable influence over their members' behaviour in certain domains of life, including matters bearing on the family economy and fertility. That influence has weakened, and its domains have narrowed, but in many societies an appreciable residue remains. Similarly, most rural societies in the past and present have local outposts of government administration, whether concerned minimally with civil order (and, often, surplus extraction) or more fully with the broad and expanding range of activities that governments promiscuously take on. The positive roles for local government lie in creating a conducive setting for economic enterprise and, perhaps coincidentally (through ensuing changes in the economics of children), for fertility decline. In addition, although potentially in conflict with these facilitative roles, local government is sometimes directly involved in programmatic efforts to raise output, alleviate poverty, and reduce fertility.

The important distinguishing characteristics of community systems for present purposes are their degree of corporateness and territoriality. The former quality governs the capacity of the group (or of an élite within it) to influence the behaviour of members to suit group interests, however those may be defined. The latter affects the likelihood that demographic behaviour will be included in the kinds of behaviour subjected to group pressure. Territoriality also facilitates orderly governance: land stays where it is.

Kinship is the chief competitor to territory in defining community systems in traditional rural societies. Natural villages may have kin ties linking many of their members, but their principal identification is as a territorial unit with more or less fixed boundaries. There are cases, however, where kinship takes precedence over territory, where clans or other kinds of corporate kingroups dominate the local-level social landscape. Bangladesh is often cited as an instance of this situation (see Bertocci 1970); Sub-Saharan Africa could provide many others. In some societies, there are several distinct bases of affiliation, no one of them dominant, each generating a system of local groupings with a particular domain of influence (see, for example, Geertz 1959 on Bali). And there may be extreme cases—exemplified apparently by the Ik of Uganda (Turnbull 1973)—where there is no significant affiliative principle.

Pre-industrial Japan and Switzerland can be taken as illustrations of strong corporate-territorial community systems. Villages in both countries could and did exact a high degree of conformity from members and tightly controlled new household formation, and they could presumably cope readily with free-rider problems. Modern examples, different from one another in many respects but not in these ways, are villages in China and South Korea—mobilized in support of government policies and programmes but still far from being mere instruments of government authority. (See the account by Parish and Whyte (1978) of the largely successful resistance by the leadership of village communities in China—the production brigades—to co-optation as local government agents.) These cases contrast strikingly with, say, villages in contemporary Nepal, which appear to be nearly helpless to enforce sound agricultural and forestry practices under rapid population growth, let alone regulate that growth itself, or in Bangladesh, where the prominent local-level organizational roles are played by kin-groups and factions rather than territorial communities. In Africa, similarly, village organization has traditionally been quite weak in comparison to tribe and lineage. There seems little indication that the gradual conversion of tribal land systems into private smallholdings is being accompanied by the strengthening of village roles.

Corporateness and territoriality are community characteristics that appear to facilitate demographic restraint even under an adverse family system. Whether they also help to promote vigorous economic performance is more questionable: indeed, a strong case can probably be made that they as often as not impede it. (The incompatibility between village-level institutional arrangements deemed necessary to meet demographic objectives and those seen as best designed to promote rapid economic growth is a problem currently bedevilling agrarian policy in China.)

The local-level organizational structures relevent to agrarian outcomes also include the lower reaches of the government administrative apparatus. The territorial basis of this apparatus can be taken for granted. An important consideration in its effectiveness, however, is the societal level down to which it operates. It is arguably impossible for government to make natural villages effective administrative units, since the face-to-face contact that characterizes them generates local pressures and loyalties strong enough to capture any officialdom at this level. On the other hand, there are problems in having the lowest-level administrative unit too far above the village. Historical evidence suggests that the administrative vacuum set up by such a gap will be quickly filled by informal political entrepreneurs, ready to act as brokers in relationships between government and governed and in the process forging a new, extra-administrative organizational system designed to maximize brokerage. Such systems grew up in the colonial period in South Asia as an outcome of the attempts by the colonial authorities to extract revenues with minimal effort. The beneficiaries of the system compose a powerful coalition that can block change. This administrative legacy has been a major impediment to rural development and achievement of lower fertility in both Bangladesh (see Arthur and McNicoll 1978) and parts of northern India.

Family–Community Configurations

The combination of family system and local community and administrative organization defines a rural institutional environment that can variously promote or impede agrarian development. Historically, the European family system was very often associated with a corporate-territorial community structure. While the resulting pattern of economic growth and demographic regulation owed something to the organizational design at each of these levels, under the prevailing low-to-moderate rate of population growth, the family was the main locus of control. If family controls on marriage were overwhelmed—supposing, say, that Europe had been faced with the pace of demographic change of the present-day developing countries—then the option of a strong community role existed. (Rapid population growth need not, of course, threaten rural welfare if there is an open land frontier, as there was in pre-industrial America, or if a substantial fraction of the rural population's natural increase is being drawn into the cities.)

In the case of the other two kinds of family systems considered, joint and African, community (and administrative) organization is altogether more significant. These family systems lack the in-built demographic regulation that the European system provided; higher-level feedbacks or government interventions are likely to be needed to curtail rapid population growth or counter its economic ill effects. In much of East and South-east Asia the institutional forms existed on which community roles in agrarian development could be based, although their potentials were somewhat obscured prior to the land reforms of the 1950s. Governments of otherwise quite divergent political orientations recognized and made use of these structures in the development effort. An effective and, at least in the economic sphere, relatively non-intrusive administrative system was another important element of success. In contrast, South Asian and Sub-Saharan African patterns of community organization seem to offer little similar scope for mobilization, even if a supportive local government system existed. Some individual families can and do prosper nevertheless in such an environment, but, without a vigorous and labour-absorbing non-agricultural economy, the institutional obstacles confronting broad-based rural development are severe.

No serious typological intent underlies this discussion. Even the brief account of rural institutional forms in the present chapter introduces further distinctions that should enter any minimally adequate attempt at classification of family–community configurations. That such classifications are feasible and would be potentially enlightening contributions to debate on population growth and agricultural development we have little doubt.

The Persistence of Institutional Forms

The family is sometimes portrayed as an institution that is relatively malleable in the face of economic or demographic change. Until quite recently, for example, a common perception held that the household in Western Europe experienced a radical transformation in the course of agricultural and industrial development, evolving from an extended to a nuclear structure during this period. While in function the family unquestionably did undergo fundamental change—as the workplace increasingly became separated from the household, for example, and as formal education came to replace some of the socialization functions of the family—in many respects the common perception was incorrect: the north-western European household appears never to have been extended in structure (or at least not for a millennium). The emerging image of the family through the centuries is one of remarkable continuity and persistence in form (Laslett and Wall 1972).

In the developing countries, a similar perception of malleability in family structure is evident in the scholarly work of the 1960s and 1970s that focused on the relationship between household structure and fertility within particular societies (Burch 1983). This research was informed largely by an anticipation that modernization in the post-war period would lead to a shift from 'traditional' extended family forms to 'modern' nuclear families. Now, of course, it is understood that the joint family system produces life-cycle periods of both nuclear and joint household residence. It is further understood that the period of joint residence is generally greater for wealthier families than it is for the poor. There is very little evidence, in fact, that post-war economic, social, or demographic change has had an appreciable impact on the prevailing systems of household formation and structure in most developing countries. Certainly, the extent of change anticipated by the research just mentioned was unwarranted.

The major features of the north-western European system of household formation—late age at marriage for both sexes, the independence of a couple after marriage, and the circulation of young unmarried people among households as servants—can be observed as early as the twelfth century and before. Analogous rules of household formation in joint systems—early marriage for men and even earlier marriage for women, residence of the newly married couple for a period in the parents' or another elder's household, and the subsequent splitting of households with several married couples—also seem impervious to change through time. Although the historical statistical record is much thinner for the joint household system, the distinctive and quite invariant distributions of household composition that the joint system produces have been shown for societies as disparate as rural India in 1950, Nepal in 1976, rural China in 1930, and fifteenth-century Tuscany (Hajnal 1982). Evidence on African family systems is scarcer still. That these also show considerable resilience, however, is suggested by the observation that Caribbean family arrangements retain discernible African characteristics that survived the trauma of slavery.

What accounts for the persistence of household formation systems? First, as noted earlier in the case of intergenerational property transfers and welfare institutions, particular family systems are associated with distinctive solutions to a variety of problems common to all societies. This functional integrity generates considerable inertial force. Second, one can discern patterns of advantage and interest associated with particular family systems that also serve as a source of resistance to change. In the joint system, the older generation might be viewed as a blocking coalition that, in order to achieve security for itself, is prepared to transfer the associated costs to younger generations. Men as a group can, in situations of male dominance, represent a similar conservative force, as can local élites. A third factor contributing to the persistence of family and kinship institutions is family morality (notions of right or wrong applied to the behavioural options governing family formation, dissolution, property devolution, and so on) and its reification in religious doctrine, ritual, and socialization practice. (Formal religion may also have a determining role in the evolution of particular family systems; Goody (1983) presents a persuasive argument to this effect in the case of Europe.)

We do not of course argue that joint or lineage-based family systems are immutable up to extremes of economic deprivation. Feedback effects from economic and demographic behaviours bring pressures to bear on family arrangements and make realities increasingly depart from culturally speci-fied ideals. That these feedbacks by and large have not been determinative under historical rates of population growth and technological change does not mean that they will be similarly ineffectual under the conditions obtain-ing today. As a case in point, we noted earlier that lineage roles in African family systems appear to be diminishing. While there is no consensus on the scope and significance of resulting changes in the family, possible outcomes include an emerging dominance of female-headed households (like those prevalent in the Caribbean) and a shift toward the nuclear family.

Like the family, community structures also tend to be stable arrangements although subject to erosion under rapid social and economic change. Family reconstitution studies, notably in England and Japan, have traced village demographic histories over centuries: individual family lines disappear, but the community is preserved. A very different illustration of stability is given by Wertheim (1964) in a description of pioneer settlement in Sumatra, Indonesia, outside the government resettlement programme. The settlers, drawn from densely populated rural Java, were confronted by vast areas of potentially cultivable forest. Far from shifting to extensive agriculture to accord with their new factor proportions, they instead constructed precise replicas of Javanese villages, resembling 'the ways of colonies of ants who by instinct know how to construct new communities, but ignore the outward factors which may endanger the existence of the community' (1964, 205).

Several factors making for this stability of community structure can be cited. In an insecure environment, especially one with high rates of mortality and debility, the family unit is too small to offer a reliable means of risk-spreading. Cultivation of the needed social ties for this purpose over a larger group becomes an essential part of an individual's survival strategy. Because this is the case for all, the reciprocal exchange relationships that are gener-ated acquire cultural validity, forming what is sometimes called a moral eco-nomy. Opting out of such a system, aside from physically departing, becomes very costly to the individual. For a few people early in the develop-ment process, and for many eventually, the economic gains to be made by severing such obligations outweigh those costs. There are also exogenous changes that tend to erode community values as contacts with urban life expand and as local labour and product markets are absorbed into the larger economy. These processes of attenuation ultimately suburbanize rural society, supplanting community institutions by others more specialized and less tied to locality.

Local administrative organization, although at first sight something changeable at will or whim by government, is another institutional char-acteristic that tends to persist. Successive governments may differ greatly in political orientation but be quite alike in their interests at the grass-roots level (in civil order, revenue-raising, and mobilization for economic growth). Administrative designs tend to gravitate to those that have worked before. Non-territorial affiliative ties (based, for example, on kin, clan, or tribe) can also be highly resilient, and in some cases they can undermine efforts to introduce 'rational' bureaucratic structures in local government or economic enterprise. Observations of this problem in Sub-Saharan African societies underlie Hyden's pessimistic views (1983) of early establishment of sound institutional bases for development in that region.

As we argued was the case with family systems, an important contributor to stability of social organizational forms is simple self-interest on the part of those who see themselves benefiting from the status quo. Part of those benefits can readily be applied to system maintenance. Powerful blocking coalitions to oppose institutional change are familiar in any development process. Where the benefits to such groups are large, as in the political brokerage arrangement mentioned above, the blocking efforts are likely to be similarly strenuous. In more developed economies the outcome may be the kind of economic sclerosis vividly described by Olson (1982). Short of radical institutional shake-ups through war or revolution or hyperinflation, the main hope of escape is that eventually the prospective benefits of new arrangements become great enough to permit the buying out of opposition.

Finally, a factor that promotes stability of organizational forms at supra-village levels is the spatial logic of markets and administration. In the former case, this logic dictates an economic structure of marketing areas, typified by the elegant hexagons of Christaller–Lösch central-place theory (see Whyte 1982). Local administrative systems, not surprisingly, often make such areas into administrative units, adding selected government functions to the economic functions of market towns, and thus transferring the stability of the one to the other. The commune in China was an administrative division roughly superimposed on the marketing area (Skinner 1965).

Agrarian Impasses and Resolutions

In this chapter, we have argued that prevailing patterns of family and community organization in most contemporary developing countries pose difficulties for successful agrarian outcomes that did not have to be faced in many historical cases. Very broad distinctions in institutional forms have been used in order to simplify exposition, but a finer-grained treatment would probably not change the essentials of the argument.

Agrarian development in the contemporary world confronts the added difficulty of having to take place under conditions of rapid population growth. Where a vigorous, labour-absorbing industrial sector exists, this growth may be of little consequence: ultimately most of the rural population will be drawn into the modern economy. For the majority of the developing countries today, any such prospect is at best distant. We have seen how some common institutional configurations can protect population growth from negative feedback effects. In turn, population growth militates against certain kinds of institutional change that could help establish such feedback. Under these circumstances the broadly defined agrarian system (encompassing family and other local institutions as variables rather than parameters) is in an impasse. Endogenous mechanisms of population-induced technical and institutional change are no match for its resilience.

What are the possible kinds of change that might resolve such an impasse and set the system on a new development path? A simple fourfold classification can cover the field. First, there is the possibility that sheer informational activity, spreading knowledge of alternative systems and behaviours and of the potential gains they have to offer, can itself effect the escape. Both demographic behaviour and the roles assigned to local government and community could be candidates for reconsideration. The case can be stated in terms of reducing the transaction costs associated with changing (as opposed to operating) an institution. Such costs include those of 'reaching agreement to make a change or, if agreement cannot be reached by negotiation, the cost of using force or other means to obtain acquiescence' (Cheung 1982, 38). In an argument concerning prospects for broad-scale emergence of private enterprise in China, Cheung maintains that the high cost of obtaining reliable information about alternative economic arrangements has been a major obstacle to this direction of institutional change. Analogus arguments might be made in the case of interest to us here.

Less ethereal than this is the prospect of escape through technological change. In a loose variant of once-popular equilibrium-trap models, a sudden surge in agricultural production associated with new technology creates a situation in which the relative costs of institutional change are quite low. This is apparently what has happened in the Indian Punjab and in other areas where the Green Revolution package of technology has most successfully taken hold. The modestly labour-using properties of this technology, combined with the dramatic productivity gains that it has engendered, have elevated many indigenous Punjabis to a *rentier* status, while drawing migrant labour from the neighbouring, less developed areas of Uttar Pradesh and Bihar. The availability of such migrant labour reduces the potential for polarization and friction within the indigenous population (as does the absence of caste among the dominant Sikh community). The broad base of development and welfare gains in this region (real wages have improved also) can generate a slack in which the economic rationales for particular institutional forms (including perhaps such fundamental characteristics as parent–child relationships) recede and there is scope for new configurations to emerge. New advances in agricultural technology since the Green Revolution show some promise of yielding opportunities of this sort in many other regions.

A third class of policy comprises what can be called minor structural change—interventions that may themselves have relatively low political costs but that may set up repercussions with larger consequences. Examples would be devolutionary shifts in systems of local finance, reallocation of government functions among centre and locality, and privatizing (or easing government monopoly) of certain kinds of service activities. Changes in the direction of tightening local controls or organizing formerly inchoate interest groups could also fall in this category.

Finally, the agrarian impasse can be resolved through radical structural change in periods in which the high social and political costs are discounted. The erratic agrarian history of modern China gives little confidence that a system created through such a process will necessarily be efficient; more promising East Asian examples are perhaps the Japanese colonial agrarian reforms in pre-war Korea and Taiwan and the post-war land reforms in these countries.

Our purpose in this chapter is not to derive policy, and the preceding paragraphs are no more than a token gesture in that direction. Rather, our concern is with expanding the domain of interest of research on agrarian change to encompass fundamental societal institutions that play a large part in determining long-run outcomes. That these institutions are singularly intractable in the face of the usual array of levers of government policy is a poor reason for neglect.

References

Arthur, W. B., and G. McNicoll (1978), 'An Analytical Survey of Population and Development in Bangladesh', *Population and Development Review* 4, 23–80.

Bartlett, H. H. (1955–61), *Fire in Relation to Primitive Agriculture*, University of Michigan Department of Botany, Ann Arbor.

Bertocci, P. J. (1970), 'Elusive Villages: Social Structure and Community Organization in Rural East Pakistan', Ph.D. dissertation, Michigan State University.

Binswanger, H. P., and J. McIntire (1984), 'Behavioural and Material Determinants of Production Relations in Land-abundant Tropical Agriculture', Discussion Paper no. A R U 17, Agricultural and Rural Development Research Unit, World Bank.

Boserup, E. (1965), *The Conditions of Agricultural Growth*, Aldine, Chicago.

Burch, T. K. (1983), 'The Impact of Forms of Families and Sexual Unions and Dissolution of Unions on Fertility', in R. A. Bulatao and R. D. Lee (eds.), *Determinants of Fertility in Developing Countries*, vol. 2, Academic Press, New York.

Cain, M. (1983), 'Fertility as an Adjustment to Risk', *Population and Development Review* 9, 688–702.

Cheung, S. N. S. (1982), *Will China Go 'Capitalist'? An Economic Analysis of Property Rights and Institutional Change*, Institute of Economic Affairs, London.

Fukutake, T. (1972), *Japanese Rural Society*, Cornell University Press, Ithaca, N Y.

Gaunt, D. (1983), 'The Property and Kin Relationships of Retired Farmers in Northern and Central Europe', in R. Wall, J. Robin, and P. Laslett (eds.), *Family Forms in Historic Europe*, Cambridge University Press, Cambridge.

Geertz, C. (1959), 'Form and Variation in Balinese Village Structure', *American Anthropologist* 61, 991–1021.

Goody, J. (1983), *The Development of the Family and Marriage in Europe*, Cambridge University Press, Cambridge.

Gourou, P. (1965), *The Tropical World: Its Social and Economic Conditions and its Future Status* (4th ed.), London.

Hajnal, J. (1982), 'Two Kinds of Preindustrial Household Formation System', *Population and Development Review* 8, 449–94.

Hanley, S. B., and K. Yamamura (1972), 'Population Trends and Economic Growth in Preindustrial Japan', in D. Glass and R. Revelle (eds.), *Population and Social Change*, Arnold, London.

Hayami, Y., and M. Kikuchi (1982), *Asian Village Economy at the Crossroads*, University of Tokyo Press, Tokyo.

Hyden, G. (1983), *No Shortcuts to Progress: African Development Management in Perspective*, University of California Press, Berkeley.

Kussmaul, A. S. (1981), *Servants in Husbandry in Early England*, Cambridge University Press, Cambridge.

Laslett, P., and R. Wall (eds.) (1972), *Household and Family in Past Time*, Cambridge University Press, Cambridge.

North, D. C., and R. P. Thomas (1971), 'The Rise and Fall of the Manorial System: A Theoretical Model', *The Journal of Economic History* 3, 777–803.

—— (1973), *The Rise of the Western World: A New Economic History*, Cambridge University Press, Cambridge.

Olson, M. (1982), *The Rise and Decline of Nations: Economic Growth, Stagflation and Social Rigidities*, Yale University Press, New Haven, Conn.

Parish, W. L., and M. K. Whyte (1978), *Village and Family in Contemporary China*, University of Chicago Press, Chicago.

Rosenberg, N. (1976), *Perspectives on Technology*, Cambridge University Press, Cambridge.

Skinner, G. W. (1965), 'Marketing and Social Structure in Rural China, Part 3', *Journal of Asian Studies* 24, 363–99.

Smith, R. M. (1984), 'Some Issues Concerning Families and their Property in Rural England 1250–1800', in R. M. Smith (ed.), *Land, Kinship and Life Cycle*, Cambridge University Press, Cambridge.

Smith, T. C. (1977), *Nakahara: Family Farming and Population in a Japanese Village, 1717–1830*, Stanford University Press, Palo Alto, Calif.

Turnbull, C. (1973), *The Mountain People*, Cape, London.

Wertheim, W. F. (1964), 'Inter-island Migration in Indonesia', in W. F. Wertheim, (ed.), *East–West Parallels*, van Hoeve, The Hague.

Whyte, R. O. (1982), *The Spatial Geography of Rural Economies*, Oxford University Press, Delhi.

6 Population Growth, Infrastructure, and Real Incomes in North India

ROBERT E. EVENSON
Economic Growth Centre, Yale University

The growth of population relative to the natural and reproducible resources of an economy is of considerable interest to policy-makers. Proponents of aggressive family-planning or population-control programmes base their support for such programmes on the proposition that families, particularly in low-income countries, face incentives that are inconsistent with economic and social optimality. A family does not take into account the 'externality' effects of an added family member in its contraceptive decision.[1] These externality effects take the form of additions to the supply of labour in labour markets. Additions to the labour force result in lower marginal and average products of labour and in lower-equilibrium wages and incomes.

A number of social scientists, however, have argued that a simplistic Malthusian interpretation of these externality effects is inconsistent with historical experience. As population grows relative to a fixed land base (as population density increases), certain changes in economic and social organizations and in technology and infrastructure may be induced by the increase in population growth. These population-induced effects may affect the Malthusian decline in the marginal and average product of labour.

An influential book by Ester Boserup (1965) makes this point forcefully. Boserup discusses the changes in agricultural organization and crop cultivation techniques that accompanies population growth. She notes that population-growth-induced changes in fallow systems from swidden and long-fallow systems to annual- and multiple-cropping sytems were accompanied by new cultivation techniques and investments in land and irrigation capital.

Presented at the IUSSP Seminar in Population, Food and Rural Development, New Delhi, December 1984. Constructive comments by A. Kelley, J. Simon, and M. Rosenzweig on an earlier version are acknowledged.

[1] Actually, many proponents of family-planning programmes argue that the problem is not externalities *per se* but lack of contraceptive knowledge and technology. They would argue that *desired* family size objectives are consistent with societal goals and that family-planning programmes enable families to achieve desired family size.

Julian Simon (1977 and 1982) is a stronger advocate than Boserup of population-induced investments and technical change. He argues that the 'Verdoorn' effect (that output demand expansion causes productivity gains) is important in historical data and that population growth induces agricultural investment by both the public and private sector (especially in irrigation), as well as in rural infrastructure. He is critical of simple calculations of the value of averted births and suggests that when population-induced effects are taken into account, moderate population growth rates may be more desirable—from a welfare point of view—than low rates.

Economists concerned with institutional change, like Hayami and Kikuchi (1981) and Roumasset (1979), also consider the possibility that population growth induces institutional change. Some of this inducement takes the form of scale economies in labour markets. Others, however, stress the role of a large supply of labourers or potential tenants as a factor reinforcing particular types of contracts and linkages among contracts. These effects may be translated into induced institutional change in the opposite direction (Bardhan and Srinivasan 1971, Braverman and Srinivasan 1984).

Actually, the proponents of the Malthusian perspective would acknowledge that many changes do accompany increases in population density. They would point out, however, that such changes do not *enlarge* the technological opportunity set of the economy. They simply cause the utilization of different components of an *existing* fixed set of technologies. Induced investments in land substitutes, particularly in irrigation capital, may occur under certain forms of organization, but such investment is limited by the low-income constraint facing high-density societies.[2]

This chapter offers an empirical perspective on the issue of population-induced effects. I use data from India for the 1959–75 period. My procedure requires four steps. I first specify a long-run analysis of the determinants of the structure of Indian agriculture (meaning by *structure* the physical, biological, economic, and technological environment that farmers face in India). Population density is specified as a determinant of this structural environment. Second, the impact of the components of the structural environment both on the supply of agricultural goods and on the demand for variable factors of agricultural production (including labour) is estimated. Then a general equilibrium model of the northern Indian agricultural sector is developed, allowing treatment of prices of agricultural goods and factors including labour as determined by the model rather than exogenously given. This permits the computation of the impact of population-induced

---

[2] Rice cultivation systems in Meiji Japan in the mid-nineteenth century and in Java in the early twentieth century could be noted as examples of Malthusian population-induced systems. In both economies, extremely high labour–land ratios were attained (more than 400 person-days per hectare in Japan and almost as much in Java) and investment did occur. None the less, these economies could be characterized as being approximately in a subsistence state of impoverishment because of population growth.

structural change (or Boserup effects) on the equilibrium prices and quantities in this model. It also allows computation of the changes in prices and quantities that ensue from the changes in demand for agricultural goods and in the supply of labour that accompany population growth. Finally, these changes in prices and quantities are converted into real-income impacts for five population groups in the Indian economy.

This empirical exercise allows one to conclude that there are indeed significant population-induced or Boserup effects in northern India. However, they only partially offset the negative consequences on real incomes that ensue from population growth effects on product demand and on labour supply. Further, it can be concluded that the deliberate investments by Indian governments and private individuals that are not population-induced have been more than sufficient to offset the net negative population growth impacts in northern Indian agriculture over the period.

The chapter is structured as follows. The next section provides a brief schematic discussion of the empirical steps involved in the empirical exercise. 'Population Density and Agricultural Structure' then discusses the logic of the empirical specification of the relationship between population density and agricultural structure. The next section is 'Estimates of Population-induced Structural Change' (step 1), and then the subsequent section reports estimates of the structure–output-supply–factor-demand relationships (step 2). Following that is a discussion of the policy model (steps 3 and 4), which reports the real-income impacts of population change and compares these with other policies. The concluding section discusses policy implications.

While the central concern of this chapter is with the effects of population pressure on agricultural development, in no way is it intended here to deny that agricultural development has an effect on population growth. However, because this chapter focuses on a relatively short post-independence period (1959–75), changes in agricultural development such as the high-yielding varietal technology introduced in the 1960s would have affected birth rates in the 1960s but would have had little effect on the numbers of agricultural workers and cultivators until some fifteen years later. Thus, this chapter sets aside the causal relationship between agricultural development and population growth itself on the grounds that a long time-lag is involved.

An Overview of the Exercise

The purpose of this section is to sketch out the theoretical foundations for the sections to follow. Figure 6.1 provides a schematic view of the approach. The core of the approach is the specification of producer behaviour, from which can be derived *supply* curves for agricultural products and *demand* curves for variable agricultural inputs. These are derived for a given *structure* of production.

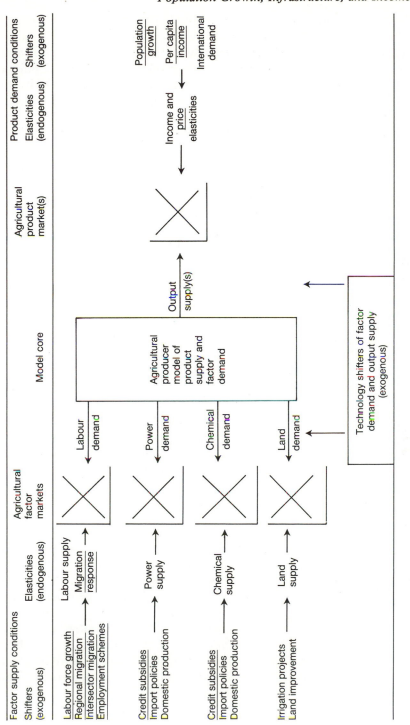

Fig. 6.1.   Distributional policy analysis model: schematic representation

Note: The core model and underlined items will receive particular emphasis in estimation and policy simulation.

The northern India study specifies a producer core in which producers supply four crops or crop combinations: rice, wheat, coarse cereals (maize, sorghum, and millet), and other crops (sugar-cane, pulses, and so forth). Producers employ four variable factors of production: fertilizer, animal power, tractors, and labour. The structure of these farms is measured by the degree of rural electrification, investment in roads, rainfall and climate, irrigation investment, size of farm, and the availability of new technology as measured by the proportion of area planted with high-yielding varieties and by past investment in research and extension.

Step one of this analysis entails specifying the relationships among the structure underlying the producer core, population density, and other variables. Step two is the estimation of the core system of equations. These equations provide the supply side of the agricultural product markets and the demand side of the agricultural factor markets. Step three completes the model by using estimates of the demand side of the product markets and the supply side of the factor markets. An equilibrium set of prices and quantities in both the factor and product markets are then obtained. One can then change one or more policy variables, including the structure variables affecting these markets, and one can analyse the impact this has on changes in equilibrium of prices and quantities.

In addition to these population-induced structural change effects, population affects the demand for agricultural outputs and the supply of labour to the sector. Equilibrium prices and quantities also change because of these effects.

The fourth step is important for the conclusion of the chapter. This entails translating the changes in prices and incomes from population-induced structural change into real-income effects for five population groups in India: small, medium, and large farmer groups, landless workers, and urban households. This is done by obtaining appropriate income and consumption weights for each group.

Population Density and Agricultural Structure

Individual economic units (farms and households) have control over a limited number of variables in the short run. They can allocate land and buildings to different agricultural enterprises. They can allocate their own labour time and their own animal-power time to different activities on the farm, in the household, and off the farm. They can purchase fertilizer and other variable inputs. They can also choose among various alternative technologies of production available to them at each time interval. In the short run, however, they cannot choose many elements of the structure in which they produce. Prices, for example, are determined by market forces, and farmers and consumers are thus 'price takers'.

It is this degree-of-control concept that is used to classify various com-

Table 6.1   Structural variables: Indian agriculture

Variable	Definition	Variable means	
		1959	1975
Type 1:	Variables determined by broad national and international economic forces		
POPDEN:	Rural population per 1,000 ha net cropped area	2,808	3,576
URBAN:	Percentage of population in urban centres	17.97	21.56
LITERACY:	Percentage of adult male literacy	20.54	28.13
GINI:	Gini rate of consumption expenditures	0.307	0.281
HYV:	Percentage of gross cropped area planted with high-yielding varieties	0.000	16.87
BORROW:	A measure of borrowable research in other states with similar geo-climates	0.599	4.282
Type 2:	Variables determined primarily by state government bodies		
ROADS:	km of roads per 1,000 km^2 arable land	1.207	2.270
ELEC:	Percentage of rural villages with electrical service	8.229	48.47
STATEREV:	Government revenues generated at the state level	4,322	47,200
BLKS:	Percentage of extension blocks in the state at stage two level or higher	14.6	99.9
EX/EXP:	Expenditures on extension service	7,547	50,046
RESEXP:	Expenditures on all agricultural research at state level	3,991	16,768
Type 3:	Variables determined partly by long-run farmer decisions		
FARMSIZE:	Cultivators/NCA (see below), a measure of average farm size	655.6	588
IRRINT:	Percentage of gross cropped area that is irrigated	21.72	29.37
NCA:	Total net cropped area in region (i.e., area planted at least once)	9,463	9,892
DLNCA:	Change in NCA over past three years (an index of cost of land expansion)	– 56.98	142.81

ponents of structure into different types. Table 6.1 summarizes these classifications and defines variables that are actually measurable at the state level in India.

Structural variables are classified into three basic types. Type 1 variables are fundamental outcomes of broad social, cultural, and economic forces. These variables are not subject to analysis here. These are treated as the fundamental determinants of other (type 2 and 3) structural variables. This set of variables is fundamental to economic growth and welfare. Population density, POPDEN, is the central variable. Note that a significant 27 per cent increase in population density occurred between 1959 and 1975, as measured in the state data set. This change in population density had important effects. It led to an increase in demand for food and other commodities. It also caused an increase in the supply of labour to labour markets in India. Less obviously, this increase in population density also affected the other structural variables in Indian agriculture.

Included also are a measure of urbanization, the Gini ratio of the distribution of consumption expenditures, and the literacy rate. These are other structural characteristics of the rural population and markets that are greatly susceptible to change by government action. Note that urbanization increased significantly over this time period, as did rural adult male literacy. The Gini ratio moved toward equality.

Two variables that form part of the technology base are classified as type 1 variables. The reason for classifying them thus is that they are essentially beyond the control of state government. High-yielding varieties (HYV) of wheat and rice were initially imported from abroad. State research programmes were quite adept at developing further adapted HYVs of Indian origin from the initial imported varieties, but the scope for this work was largely determined by the original pattern of imported technology.

Also included is a measure of research intensity in other states with sufficiently similar geo-climate characteristic, so that their research programmes might be expected to produce importable technology. States may respond to this importable or borrowable technology by *free-riding* (that is, investing less in their indigenous technology base). Conversely, they may respond positively to the adaptable research opportunities afforded by such technology. This issue is examine in the section to follow.

Type 2 structural variables basically represent political decisions by state government bodies. They include investment in the technology base (research and extension) as well as in other infrastructure–institutional variables. Several of these variables can be viewed as having an investment dimension and a level or capital stock dimension. For example, investments can be made in road building and extending electric power to new villages to add to the stock of road capital and electricity-generating capital. Similarly, investments in research (and to a lesser extent extension) contribute to the technology capital available to farmers.

It is of interest to note that Indian states have in fact expanded investments in type 2 variables significantly over the 1959–75 period. Research expenditures expanded fourfold, and the stock of relevant research (measured by LOCAL) increased eightfold. Extension spending increased more than sixfold. Road expansion and rural electrification expansion was also significant.

While it is not the central topic of this chapter, it is worth noting that the general investment in type 2 structural change by Indian government units has had important economic and social consequences. It is this investment that staved off the Malthusian outcomes inherent in the increasing population density in India over the period. Indeed, as can be seen in the next section, population density itself had something to do with this investment pattern.

The final set of structural variables, type 3, have a significant component of farmer decision-making associated with them. Irrigation intensity (IRRINT) might be regarded as a type 2 variable in areas where large-scale irrigation projects dominate. In areas where tube wells and small-scale projects dominate, farmers and groups of farmers at local government levels determine investment in irrigation. Similarly, the expansion of net cropped area may also have some state government investment.

Farm size is more directly the result of farmer decisions. It is a long-run outcome of essentially the same maximizing behaviour that determines short-run variable output and input decisions by farmers. Because farmers take prices as given, prices could be considered to be determinants of this behaviour. It is also reasonable to postulate that state governments respond to price regimes in setting their investment priorities. However, because prices themselves are changed by population growth (see the section on conclusions) they are not included in this analysis.

Estimates of Population-induced Structural Change

The logic of the aforementioned typology also underlies the empirical specification utilized in this section. Two concepts characterize the specifications. The first is that the decision-making unit in a position to alter the structural variable is responding to factors not directly under its control. The second is that of recursiveness—that the decision-making unit is currently responding to past decisions. Type 1 structural variables are treated as exogenous determinants of type 2 and 3 variables on these grounds. Type 1 variables are determined by international events and long-run processes at the national level. Type 2 variables, on the other hand, are determined primarily at the state level and are the outcomes of political processes. Type 3 variables are determined by farmers themselves.

The analysis of type 2 variables is undertaken with state-level data for 14 major states for the period 1959 to 1975. This is in keeping with the decision-making units involved. I do not have adequate farm-level data to pursue the analysis of type 3 variables. I do, however, use district-level data in the analysis to approximate the more micro-level data desired. These data are for the districts in four northern states, Punjab, Haryana, Uttar Pradesh, and Bihar, for the 1959–75 period.

In order to strengthen the natural recursiveness of the specification, all dependent variables are in current flow or investment terms where relevant. For example, research spending is in terms of current expenditures, rural electrification is measured as the change in percentage of villages with electricity, and so forth. Independent variables are in stock terms, and they

Table 6.2   Determinants of public sector investments (Type 2) Indian agriculture estimated elasticities

Determinants	Research expenditures	Extension expenditures	Changes in roads	Changes in electrification	Provision of agricultural credit
Population density	-1.46^b	0.10	0.02	-0.03^b	-0.08^b
Consumption Gini	1.29^a	-0.59	0.09	-0.06	0.06
Urbanization ratio	-0.02	-0.19^a	0.07	0.02^b	0.06
Literacy level	1.95^b	2.05^b	0.01	0.12	0.21^b
HYVs	-0.20^b	-0.01	-0.01	-0.02	0.04^b
Borrowable research	0.83^b	0.52^b	0.02^a	0.04	0.01
Net cropped area	0.71^b	1.17^b	-0.28^b	-0.01	-0.01^a
Change in net cropped area	0.02	0.00	0.00	0.03	0.04^b
R^2	0.583	0.719	0.182	0.044	0.768
F	3.07^a	56.4	4.9	1.9	73.2

[a] Asymptotic T > 1.5 and < 2.0.
[b] Asymptotic T > 2.0.

are lagged. For example, when rural electrification is an independent variable, it is measured as the average percentage of villages electrified in the four years prior to the date of the observation.

Table 6.2 reports the estimated elasticities (evaluated at the mean of the data, and so on) of determinants of the type 2 variable. The estimates are based on simple linear ordinary least squares (OLS) estimates (R^2 and F ratios are reported; note that the rural electrification equation has little explanatory power). Interestingly, population density has negative impact on research investment, rural electrification, and credit provisions. The positive effects on extension expenditures and roads are not significant. Literacy emerges as the key variable stimulating type 2 investments. This analysis has not attempted to determine whether literacy itself is affected by population density. Population density would be expected to have a negative effect on literacy.

It may be note further that the imported Green Revolution HYVs reduced state research expenditures, but they did stimulate credit provision. Borrowable research, on the other hand, stimulates *both* research *and* extension spending.

Table 6.3 reports the estimated elasticities for type 3 variables. The specification of these equations includes population density, consumption Gini, adult male literacy levels, and HYV imports as in the specification of type 2 variables. In addition, it includes the lagged stocks of the type 2 variables. The basis for including these stocks is that they constitute the public sector infrastructure that farmers and local decision-makers face. Because population density and other variables also influence these stocks, their interpretation in Table 6.3 is that they have additional or independent influence on long-run farm behaviour.

Table 6.3   Determinants of farm structure (type 3) estimated elasticities[a]

	Farm size	Irrigation intensity	Net cropped area
Population density	− 0.24[b]	0.61[c]	0.67[c]
Consumption Gini	1.25[c]	—	1.66[b]
Literacy level	0.87[c]	—	5.00[c]
High-yielding varieties	− 0.10[c]	− 0.24[c]	0.11[c]
Research intensity	0.47[c]	0.32[c]	0.11[c]
Extension intensity	− 0.21[c]	0.32[c]	—
Roads	0.08	2.51[c]	—
Rural electrification	0.01	0.69[c]	—
Credit provision	− 0.71	− 0.01	—
R^2	0.671	0.562	0.550
F	82.7	51.1	45.23

[a] All values are elasticities computed at sample means.
[b] Asymptotic T > 1.5 and < 2.0.
[c] Asymptotic T > 2.0.

Population density emerges as having strong impact independent of the type 2 investments. It stimulates a reduction in average farm size and significant expansion in irrigation intensities and in net cropped area. Literacy stimulates larger farm size and expansion in net cropped area. Past investments in research stimulate increases in farm size, irrigation intensity, and net cropped area. Past investments in extension appear to stimulate irrigation investment and to reduce farm size.

Structure, Farm Output Supply, and Factor Demand

The next step in this exercise is to relate structure to output supply and to factor demand. This is done by estimating a producer core model with four output-supply equations and four factor-demand equations.

The theoretical basis for deriving the core relationships rests on the duality between maximized profits and the technical transformation function relating the four variable outputs to the four variable factors and the structure variables. When certain restrictions hold for the maximized profits function, duality theory ensures that they also hold for the transformation function. The important thing about this fact is that one can specify a functional form for the maximized-profits function directly. This is much simpler than specifying a functional form for the transformation function and then solving for the maximized-profits function. The maximized-profits function does not include choice or endogenous variables such as the variable products or variable factors of production. They are eliminated from the expression by substitution of first-order conditions. Thus, maximized profits are expressed as a function of prices and structure variables. One can then apply the Hotelling–Shephard lemma to the maximized-profits

Table 6.4   Variables dictionary: North Indian data set observations on 22 regions, 1959–74.

	Means		
Variable definitions	Wheat region	Rice region	All
1. Variable farm outputs			
Wheat	20,678.19	10,124.76	16,360.88
Rice	4,319.35	22,083.35	11,586.44
Cereal grains	5,660.20	4,467.20	5,172.10
Other crops	25,833.99	16,114.35	21,857.78
2. Variable farm inputs			
Labour	22,006.25	41,818.24	30,111.16
Animal power	21,841.10	50,139.95	33,417.90
Tractor services	1,038.04	256.74	718.42
Fertilizer	4,155.17	2,641.76	3,536.05
3. Prices			
Wheat	2.215	2.291	2.246
Rice	2.058	1.879	1.984
Cereal grains	2.174	2.390	2.262
Other crops	2.898	3.288	3.058
Labour	2.041	2.111	2.070
Animal power	1.790	1.371	1.619
Tractor services	1.577	1.577	1.577
Fertilizer	1.278	1.307	1.290
4. Structure variables			
Rural electrification (percentage of villages electrified)	38.99	15.25	29.28
Roads (km of roads per 10 km^2)	2.08	1.11	1.68
Research expenditures (cumulative expenditures, 1955 to t-2)	9.56	4.61	7.54
Research intensity (net cropped area)	1.49	0.865	1.23
High yielding varieties (percentage of gross cropped area under high-yielding varieties of rice, wheat, and maize)	10.79	7.09	9.27
Irrigation intensity (percentage of gross cropped area irrigated)	40.97	25.31	34.57
Net cropped area (thousand ha)	1,299.03	1,711.52	1,467.78
Farm size (net cropped area/number of cultivators)	0.0017	0.0012	0.0015
Agricultural labourers/cultivators	0.265	0.430	0.332
Literacy (percentage of rural males who are literate)	25.80	27.13	26.34

function. This lemma states that the first partial derivatives of the maximized-profits function with respect to an output or factor price are the output-supply and factor-demand functions. Thus, by taking eight partial derivatives, we end up with a system of four output-supply functions and four factor-demand equations. Each equation relates the quantity supplied (or demanded) to the eight variable prices and the structure variables.

The estimates were obtained using district-level data from the states of Punjab, Haryana, Uttar Pradesh, and Bihar. These districts can be group into two major groups, a primarily wheat-producing area (Punjab, Haryana, and western Uttar Pradesh) and a primarily rice-producing area (eastern Uttar Pradesh and Bihar). Table 6.4 provides a variables dictionary for the data set and reports means for the two areas. A brief description of the definition of each variable is provided. The variables are also classified as variable farm outputs, variable farm inputs, prices, and structure variables. In this analysis, it is presumed that farmers maximize variable profits by choosing the appropriate mix of variable farm outputs and variable farm inputs. These variables are thus choice or endogenous variables. I am assuming in this model that the typical farm has no control over farm prices. Table 6.5 reports elasticities computed at the means of the data for the eight equations. The two regions are pooled in obtaining these estimates. (Coefficients estimated separately for each group did not differ significantly from the pooled estimates.) By reading down each column, one obtains the elasticity effects of each price and structure variable on the output-supply or input-demand variable in question. For example, in the first column one can see the estimated effects on wheat supply of the wheat price, the rice price, and so on, all the way down to the research variable (all statistically significant variables are indicated by letters). The wheat-supply elasticity with respect to its own price is 0.370. This means that a 10 per cent increase in the wheat price, holding all other prices and structural variables constant, will cause a 3.7 per cent increase in the supply of wheat. One can also see the consequences of a wheat price increase, holding everything else constant, not only on wheat supply but also on the supply of rice, coarse cereals, and other crops, and on the demand for fertilizer, bullock labour, tractors, and (human) labour by reading across the wheat price row in the table. Thus a 10 per cent increase in the price of wheat causes a 3.7 per cent increase in the quantity of wheat supplied, a 2.07 per cent decrease in the quantity of rice supplied, a 2.24 per cent increase in the quantity of coarse cereals supplied, and so on.

The a priori expectations of the sign and magnitude of these elasticities are borne out by these data, with only one exception. All own-price elasticities of supply are expected to be positive, and they are. The own-price elasticities of demand for inputs are expected to be negative; they are with the exception of fertilizer demand. The cross-price effects are generally as expected. In the output block, when a cross-price elasticity is negative, it means that the crops are 'substitutes'. For example, wheat and rice are good substitutes, as are 'other crops' and 'coarse cereals'. Wheat and coarse cereals are 'complements' (that is, an increase in the price of one of the pair induces an increase in the supply of both). This can happen when the two crops fit well together in an annual rotation.

Within the input block, negative cross-elasticities indicate that the input

Table 6.5   Elasticity estimates: North Indian district data set 1959–75.

Elasticity with respect to	Elasticities of output supply				Elasticities of input demand			
	Wheat	Rice	Coarse cereals	Other crops	Fertilizer	Bullock labour	Tractors	Labour
Wheat price	0.370^b	-0.207^b	0.224^a	-0.031	-0.007	0.016	0.010	0.001
Rice price	-0.128^b	0.392^b	-0.076	-0.030	-0.198^a	0.008	-0.051	-0.060^b
Coarse cereal prices	0.073^a	-0.040	0.040	-0.040^a	-0.155	-0.005	0.112	0.093^b
Other crops price	-0.058	-0.090	-0.227^a	0.176^b	0.348^b	0.006	-0.016	0.011
Fertilizer price	0.001	0.042^a	0.062	-0.024^b	0.195^a	-0.038^b	0.160	0.122^b
Bullock price	-0.025	-0.019	0.025	0.005	-0.440^b	-0.010	-0.010	0.048^b
Tractor price	0.001	0.003	-0.011	0.001	0.038	-0.001	-0.084	-0.155^b
Labour price	-0.232^b	-0.079	-0.038	-0.046	0.217^b	0.023^b	-0.103	-0.061^b
Electrification	-0.025	0.011	0.057^a	0.084^b	0.245^b	0.006^b	0.034	-0.026^b
Roads	-0.110	-0.465^b	0.373^b	-0.362^b	-0.325^b	-0.086^b	0.291^a	0.029
Rainfall	0.161^b	0.407^b	-0.173^a	0.019	0.456	0.012^a	0.208^a	0.055^b
Irrigation intensity	1.123^b	0.271^a	0.919^b	0.276^b	1.203^b	0.056^b	1.851^b	0.117^b
Net cropped area	-0.139	1.485^b	1.048^b	0.609^b	0.289	-0.022	-1.266^b	0.042
Farm size	0.224^a	0.379^b	-0.027	-0.210^b	-0.744^b	0.060^b	0.693^b	-0.285^b
HYVs	0.278^b	0.109^b	-0.074^b	-0.128^b	0.259^b	0.012^a	-0.122^b	0.030^a
Indian research	0.023	-0.085^b	-0.102^b	0.176^b	0.249^b	-0.002	0.537^b	-0.084^b

[a] Asymptotic T < 2.0 and > 1.5.
[b] Asymptotic T > 2.0.

pairs are substitutes. The data show that fertilizer may be a complement both to labour and to tractors.

Perhaps of most interest, however, are the 'structure' shifters. The results indicate that rural electrification biases the output mix in favour of coarse cereals and other crops. It biases input demand in favour of fertilizer and against labour. Roads, on the other hand, create biases in favour of coarse cereals and against other crops and are biased against fertilizer and in favour of tractor demand. It should be noted, however, that this variable, and perhaps others, may be reflecting geographical factors. For example, the rainfall variable is a strictly geo-climate variable and definitely not subject to policy modification.

Irrigation intensity and net cropped area, on the other hand, are subject to policy manipulation. Increasing irrigation investment increases all outputs and inputs, but it is quite clearly biased toward wheat and coarse cereals on the output side and toward fertilizer and tractor use on the input side. As net cropped area in the typical district expands, holding farm size constant, it is biased in favour of rice and coarse cereals and against wheat. It increases the demand for fertilizer, but decreases the demand for tractors. An increase in average farm size, holding total net cropped area constant, on the other hand, is biased against rice and favours other crops and fertilizer and tractor demand. It is biased against labour employment. Conversely, a decrease in farm size would reduce the demand for fertilizer and tractors and increase the demand for labour.

Much has been written about the Green Revolution and general technical advances in India. The general presumption of much of the literature is that the introduction of HYVs did not have biases on the input side, though it was clearly biased in favour of wheat and rice on the output side. The HYV variable clearly confirms the bias in favour of wheat and rice. It also quite clearly shows that when HYV wheat and rice are made available, the supply of coarse cereals and other crops is reduced. The results also show a bias in favour of fertilizer on the input side.

The Indian agricultural research system, on the other hand, has a strong bias in favour of other crops. It also appears to have quite strong biases on the input side. It produces technology that is reliant on fertilizer and tractors and is labour-saving.

The elasticities of total output, total variable input, and total variable factor productivity with respect to the policy-related structural variables are reported in Table 6.6.

It would appear that rural electrification has a modest effect on productivity, whereas irrigation intensity has a substantial effect. An expansion of net cropped area, holding irrigation intensity constant, has an effect approximately the same as the irrigation intensity effect. This implies that an expansion of irrigated land has an effect on production approximately twice that of an expansion of unirrigated land. The farm size variable is showing what

Table 6.6   Elasticities of total output, total input, and total variable factor productivity

	Total output	Total variable input	Total variable productivity
Electrification	0.03357	− 0.00859	0.02450
Irrigation intensity	0.58752	0.23477	0.35275
Net cropped area	0.61071	− 0.04438	0.65509
Farm size	0.06043	− 0.16189	0.22232
HYVs	0.04796	0.02265	0.02531
Indian research	0.04929	− 0.02037	0.06966

Source: Computed from Table 6.5.

appears to be very substantial variable factor scale economies. Expanding farm size by 10 per cent increases output and decreases inputs, holding total net cropped area constant. This is a large-scale effect and somewhat at odds with most interpretations of Indian agricultural facts.

This analysis implied very large returns to investment in Indian research. Because this analysis does not include an extension variable, it is probably reasonable to suppose that the research variable is picking up both a research and an extension effect. Spending on research and extension combined represented approximately 0.70 per cent of the value of agricultural product in the early 1960s. A 10 per cent increase of this spending (0.07 per cent of output) is estimated to produce a 0.7 per cent increment to net output. This can be converted to a rate of return by noting that an expenditure of 0.07 per cent of output generates a stream of net output amounting to 0.7 per cent of output after seven years. This implies an internal rate of return to this investment of 72 per cent.

Table 6.7   Population-induced shifts in output-supply and factor-demand elasticities of population-induced structure shifts:

Output supply		Input demand	
Wheat	0.531	Fertilizer	1.033
Rice	1.080	Bullock labour	0.004
Coarse cereal	1.300	Tractor	− 0.008
Other crop	0.575	Labour (human)	0.189
Total crop	0.670	Variable factor	0.174

The population density effects measured in Tables 6.2 and 6.3 can now be traced through Table 6.5 to obtain population-induced shifts in output supply and variable factor demand. Table 6.7 reports these population-induced structure effects in terms of elasticities. It is clear that they are important. An expansion of population density induces changes in structure that have quite large output effects, a 10 per cent expansion in population density inducing structural changes that produce a 6.7 per cent increase in output and the same changes inducing a 1.74 per cent change in variable input use. Of course, the changes in structure are not costless. Irrigation, expansion of area cultivated, research, and other public investment require real resources. It appears, however, that the Boserupian perspective on change is supported by the data. These induced effects, however, are not sufficiently large by themselves to prevent production per capita from declining when population expands.

Long-run Economic Effects of a Reduction in Population Growth

One can now consider the economic effects of a change in population. As noted earlier with reference to Figure 6.1, these are of three types. (1) Population affects demand. Other things equal, a 10 per cent decrease in population shifts all demand curves to the left by 10 per cent. (2) Population also affects the supply of labour. If we take a long-run perspective and presume that the ratio of actively employed workers to total population is constant and that the rural–urban employment ratio is also constant, a 10 per cent decrease in population translates into a 10 per cent leftward shift in the labour supply function. (3) Population induces structural effects.

The analysis of the preceding sections identified only the producer core responses (that is, the output-supply and factor-demand responses of profit-maximizing farmers). In order to analyse the full effects of population, one must model the complete markets depicted in Figure 6.1.

In order to complete the other side of the product markets, a demand for product structure is required. Similarly, in order to complete the missing side of the factor markets, factor-supply relationships are required. When these markets are completely modelled, prices themselves can be treated as endogenous to the model (that is, determined in the markets). With a specification of the complete markets, it is possible to calculate the effect of a resource base change (say an increase in irrigation) not only on the product-supply and factor-demand functions, but also on the equilibrium prices and quantities in each market.

The demand side of the product markets in India has been estimated by Binswanger, Quizón, and Swany (1984). Using methods similar to those used in estimating the producer core, a system of demand equations was used to estimate demand relationships. The numbers of consumers (in other words, population), as well as incomes and prices, determine aggregate demand in the product markets. Because the markets as depicted in Figure 6.1 show quantities demanded as functions of price only, population, income, international demand, industrial organization in the post-harvest industries, and transport and transactions costs are termed 'shifters' for these markets. A reduction in population, for example, shifts the demand curves to the left. This will then result in a decline in prices if no other shifters, including structure components, change. Of course, as a practical matter, when population changes, the supply of labour to the labour market will also change. Population is thus a shifter variable in the labour market as well.

The specification of the labour market is based on a study by Dhar (1980) and takes into account not only the effects of population change but also of migration among regions and sectors as well. Thus, when rural wages decline relative to urban wages, workers will migrate from rural to urban jobs (Dhar has estimated these migration responses). The urban-based demand for labour thus becomes a shifter in the rural labour market. A reduction in population growth among urban families can thus have an effect on rural families through rural–urban migration.

The supply side of the animal-power market is closely related to the supply of feed and is specified as a value-weighted aggregate of the crop supply estimates in Table 6.5. Fertilizer and tractor supply elasticities are set at 4.0, reflecting international supply through imports.

The completion of the specifications of the eight markets allows one to calculate the effects of a large number of shifters on equilibrium prices and quantities in each market. Rural incomes are determined by payments to labour and other owned factors such as bullocks, less payments to purchased chemical and power factors, plus a residual rent to fixed resources (in this case land). Consequently, changes in rural incomes can be inferred from changes in prices and quantities in the seven markets depicted. Furthermore, by adjusting for price changes, a price deflator can be constructed to convert nominal income changes to real income changes.

Quizón and Binswanger (1983) have grouped the Indian population into five groups:

1. Landless and near-landless rural households with less than one acre of operated land
2. Small farmers with one to five acres of operated land
3. Medium farmers with five to 15 acres of operated land
4. Large farmers with more than 15 acres of operated land
5. Urban households.

For each group, consumption weights and income weights were determined. Consumption weights showing the shares of the four agricultural products (and non-agricultural products in the typical consumption basket) were computed. Income weights based on the shares of income from agricultural labour, animal power, land rent, and non-agricultural labour in each group's income were also computed.

With this information, it is possible to translate changes in equilibrium quantities in the eight markets (plus residual land rents) into changes in nominal and real incomes per capita and cereal grains consumption per capita for each of the five groups. A reduction in population then can be seen in this short-run model as a combination of a demand shifter (a leftward shift in the product demand curves) and a labour supply shifter (a leftward shift in the agricultural labour supply curve). The demand shifts will have ramifications in the factor markets because product prices will fall, causing a leftward shift in the derived demand for factors. Conversely, the labour-force supply shift will have ramifications in the product markets, because rising wages will cause shifts in product supply functions.

This model can be solved for an initial equilibrium in all eight markets. The model includes an aggregate non-agricultural good being consumed and produced. This equilibrium can be expressed in rate-of-change form (that is, all equations are differentiated with respect to time) as a system of eight equations. This equation system can be expressed as

$$G\,U^1 = K^*$$

where G is a matrix of elasticities, U^1 a vector of equilibrium rates of changes in exogenous variables (prices and quantities), and K^* a vector of shifter-type variables. The effects of shifters on rates of change in endogenous variables can be solved as:[3]

$$U^1 = G^{-1}K^*$$

Table 6.8 reports the effects of a simulated 10 per cent decrease in population growth on the endogenous variables of the system, including real-income effects for the five population groups. The induced-structure or Boserupian effects are calculated separately.[4] In addition, the effects of a 10 per cent increase in the technology base (that is, a 10 per cent increase in the expenditures on research and extension and HYV in production), in irrigation intensity, and in areas under cultivation are shown for comparative purposes.

The simulation should be interpreted as *short-run* consequences of population change, technology, irrigation, and land investment. The basic elasticities in the model are estimated with methods that do not attempt to distinguish between short-run and long-run effects, and, given their nature, it seems reasonable to interpret them as short-run elasticities.

Five sets of population effects have been calculated. The first is termed a Malthusian calculation. In this simulation, it is supposed that over an extended period policies are put in place that reduce population *and* labour force growth such that at the end of the period both the size of the population and the labour force would be 10 per cent lower than in the absence of

---

[3] Note that this procedure is valid only for small changes. This is not a computable general equilibrium model where a new equilibrium is computed. This model provides impact multipliers for small changes in exogenous variables. Also see Quizón and Binswanger (1983) for a discussion of convexity of the G matrix and minor adjustments made to achieve convexity.

[4] The term 'Boserupian effects' here refers to a narrower definition of population-induced changes than Boserup uses.

Table 6.8 Simulated economic effects of population growth decline, technology investment, land investment, and irrigation investment in North India

Effect on	10 per cent decline in population					10 per cent increase in		
	Malthusian	Boserupian	Total	Rural landless only	Urban only	Technology base	Land base	Irrigation base
Real per capital income								
(a) All groups	7.77	−2.80	4.97	2.18	0.83	0.26	2.60	1.78
(b) Rural landless households	14.72	−8.36	6.36	7.68	1.69	1.12	6.64	6.58
(c) Small farm households	11.82	−0.59	11.29	3.31	−0.15	1.10	−0.11	−0.30
(d) Medium farm households	6.78	0.39	7.17	0.73	−1.11	−1.35	−3.19	−4.18
(e) Larger farm households	0.69	1.44	2.13	−1.93	−13.45	−3.54	−11.26	−13.52
(f) Urban households	7.06	−1.47	5.59	1.06	10.24	3.36	13.01	12.52
Agricultural employment	−4.80	−2.95	−7.75	−1.95	−2.29	−0.44	−0.30	−0.07
Real agricultural wages	12.94	0.38	13.32	7.33	−0.66	0.22	−1.88	−0.10
Real land rent	−25.18	40.26	15.08	−5.42	−7.40	−10.20	−31.45	−38.15

the policies. The simulation thus takes into account the reduction in demand for products and in the supply of labour.

This first calculation is of considerable interest because it shows that the effects of these policies are large and *progressive* in terms of their distribution. Real incomes of the population at large rise by 7.77 per cent. For the poorest group—landless labourers—real incomes rise by 14.72 per cent, whereas for the relatively high-income, large farmers, real incomes do not rise appreciably. The 10 per cent reduction in labour supply produces a 4.8 per cent reduction in agricultural employment and a 12.84 per cent rise in real wages. Real land rents (calculated as a residual in this model) actually fall by 25.18 per cent. It is this rise in real wage and decline in returns to land-holdings that produce most of the progressiveness in the real-income consequences.

The second column gives the simulated Boserupian effects associated with the decline in population. Because population density is lower, population-density-induced effects (Table 6.7) are lost. When they are considered, the gain in real income for the population as a whole falls by 2.8 per cent, so that the net gains are 4.97 per cent. The Boserupian effects are themselves progressive in nature: in other words, an increase in population density induces investments that favour the poor. Their loss is thus regressive. In these calculations, their loss reduces the 14.72 per cent gain by the landless by 8.36 per cent, leaving a net gain of 6.36 per cent. After adjusting for Boserup effects, however, a decline in population still has important and progressive effects. The rural landless and small farmers still gain most.

Two rather specialized population growth effects have also been simulated in columns 4 and 5 in the table. In these, the population of a particular group is simply reduced by 10 per cent (there are no Boserup effects). One way to visualize this simulation is to interpret it as a reduction in the population and labour force due to a labour recruitment programme for work in Middle Eastern countries. Column 4 shows that if this recruitment is directed only to the landless agricultural worker group it has a large and progressive effect on real incomes. The landless agricultural worker group gains more from this specialized effect than from the more general population reduction. (Actually, if only workers were recruited while families were left behind, real wages would rise even more.)

Column 5 shows the effect of recruitment from the urban population only. Here, the effect on real incomes is smaller but is probably still progressive. (The calculation assumes that when population is reduced, per capita income remains constant.) For comparative purposes, the effects of investments in technology, land expansion, and irrigation investment have also been calculated. These simulations do not measure Boserupian effects. These can be looked upon as policy options available as alternatives to population policy. Each option has very different costs, and these costs are not considered in the simulation. For example, a 10 per cent increase in the technology base (the HYV–Research stock) is much less costly than a 10 per cent expansion in the land or irrigation stock (in fact, only about ¹/₄₀ as costly).

Interestingly, all three forms of investment have similar effects. They lower food prices, raise real wages, and reduce land rents (note that these land rents do not include rents to new land or irrigation; we are presuming public ownership of these rents). Urban consumers benefit most from these programmes, and large farmers lose most (provided they do not collect newly created rents from the investment).

Concluding Comments

The relationship between population and development is complex. This chapter attempted to measure a major part of this relationship and has ignored or set aside another part. It has produced evidence that population growth has important effects not only through the demand for goods and the supply of labour but also through induced structure or Boserupian effects as well. The simulations reported show these effects to be important. In some sense, one can say that the Boserupian effects constitute something of an antidote to the negative and regressive effects of population growth on real income. Our simulations show that real incomes will fall less when Boserupian effects occur, and that the declines will be less regressive as well.

The approach taken in this chapter has also attempted to look at the role of non-Boserupian policy effects. It has shown that policy-makers can invest in technology, land expansion, irrigation, schooling, electrification, and suchlike, and that they can offset the negative effects of population growth if they choose to do so. India has in fact chosen to do so, as have most other countries, and as a consequence real incomes have not fallen since the mid-1960s. The simulation model is useful in providing a basis for comparing the costs of alternative policies designed to achieve real-income objectives.

The data were not adequate to make a full comparison between the costs of achieving a real-income goal through population policies or investments in technology and irrigation. The simulations do make it clear that a given real-income objective can be achieved at much lower cost through technology-base investment than through land expansion or irrigation investment. It would appear likely that an effective family-planning programme may achieve these goals at an even lower cost.

The chapter has concentrated on the impact of population on economic development outcomes. It has set aside the reverse relationship (the impact of development on population growth) primarily for reasons of tractability. Certainly, the reverse relationship exists. As wages, prices, and real incomes change, fertility changes in response. In general, the bias from neglecting this in the simulations offered here runs in the direction of conservatism regarding the policy implications of the measures reported here. A policy to intervene actively in household decision-making so as to achieve a reduction in fertility (through education, subsidized contraception, and similar) produces increases in real wages and real incomes in our model. These increases in real wages will produce further incentives to reduce fertility if they are not biased against women. A rise in the real value of time for women reduces fertility, while a rise in the real value of time for children has an opposite effect. On balance, I would expect that the second-round effects of a fertility reduction programme would reinforce the first-round effects, thus lowering the costs of obtaining a fertility reduction.

Finally, it merits repeating that even though there are significant population-induced effects that offset some of the deleterious consequences of population growth, the net effect of population growth in northern India is to reduce real incomes. In reaching this conclusion, I have considered most of the arguments for population-induced effects. It is conceivable that further research on the topic may alter this conclusion, but I doubt very much that it will. These findings for northern India will not necessarily hold for other regions of the world where demand conditions differ.

References

Bardhan, P. K., and T. N. Srinivasan (1971), 'Cropsharing Tenancy in Agriculture: A Theoretical and Empirical Analysis, *American Economic Review* 51.

Binswanger, H. P., J. B. Quizón, and G. Swany (1984), 'The Demand for Food and Foodgrain Quality in India', *Indian Economic Journal* 31(4).

Boserup, E. (1965), *The Conditions of Agricultural Growth*, Aldine, Chicago.

Braverman, A., and T. N. Srinivasan (1984), 'Agrarian Reforms in Developing Rural Economies Characterized by Interlinked Credit and Tenancy Markets', in H. P. Binswanger and M. R. Rosenzweig (eds.), *Contractual Arrangements, Employment and Wages in Rural Labor Markets in Asia*, Yale University Press, New Haven, Conn.

Dhar, S. (1980), 'An Analysis of Internal Migration in India', Ph. D. dissertation, Yale University.

Hayami, Y., and M. Kikuchi (1981), *Asian Village Economy at the Crossroads*, University of Tokyo Press, Tokyo, and Johns Hopkins University Press, Baltimore.

Quizón, J. B., and H. P. Binswanger (1983), 'The Distribution of Agricultural Incomes in North Indian Agriculture: A Model', mimeo, World Bank.

Roumasset, J. (1979), 'Sharecropping, Production Externalities, and the Theory of Contracts', *American Journal of Agriculture Economics* 61.

Simon, J. (1977), *The Economics of Population Growth*, Princeton University Press, Princeton, N J.

—— (1982), *The Ultimate Resource*, Princeton University Press, Princeton, N J.

Part III

Population Growth and Rural Labour

7 Population Growth and Access to Land
An Asian Perspective

AZIZUR R. KHAN
Country Policy Department, World Bank, Washington, DC

In a predominantly agrarian economy, the critical role of the initial demographic conditions in determining the pattern of development has long been recognized.[1] In these countries, agriculture provides much of the employment and income, and land is the most important productive asset. Given the initial endowment of this resource—measured in terms of the amount of land per labourer employed in agriculture or the amount of land per person dependent on agriculture for income and/or consumption—the rate of population growth can have an overwhelmingly powerful effect on the entire course and pattern of development.

One could trace the effect of rapid population growth in what might be regarded as an archetypical case in the following steps. An acceleration in the rate of population growth leads to a rapid increase in the size of the labour force. The amount of land either increases only very slowly or not at all. Industrialization, broadly defined to include growth in secondary and modern tertiary sectors, is unable to absorb more than a small portion of the increase in the labour force. The amount of land per worker employed in agriculture declines. The amount of land per person dependent on agriculture declines even more rapidly, due to an increase in the dependency ratio caused by the acceleration in the rate of population growth. Even if the effect is uniform on groups of all sizes, the result is an increased incidence of landlessness and near-landlessness. In reality, the effect of a sharp reduction in land–person ratio is often to push a disproportionate number of those at the bottom end of the distribution of land ownership (or land-holding) out of any access to land. The changed supply–demand balance for wage labour and the increased factor share of land in net output exert a downward pressure on real wages. The result is increased impoverishment of large masses of people. This is usually not enough to put a brake on demographic expansion, because of a number of factors (such as somewhat independent provision of health services, eradication of epidemic diseases and access to safe water,

---

[1] Shigeru Ishikawa's seminal work (1967) emphasizes this factor as an important element of the initial conditions that make the development pattern of Asia different from that of the industrialized countries during their development period.

and the tendency of basic nutrition to decline less sharply than aggregate income and consumption).

Looking at the Asian experience since the mid-1950s, one finds that development in a number of countries conforms to this typical pattern. However, in a number of other cases the experience has been substantially different. Reasons behind these different experiences are not hard to recognize. In the preceding description of the archetypical case, assumptions were made about the behaviour of a number of offsetting factors that are extreme and by no means universal in Asia. Thus, despite rapid population growth, a decline in the amount of land per worker can be avoided if the supply of land increases and/or rapid industrialization leads to an adequate rate of labour absorption outside agriculture. Also, a reduction in land per worker in agriculture (or land per person dependent on agriculture) need not lead to increased landlessness if ways are found for increased productive absorption of labour on land and/or redistributive institutional reforms are brought about. Moreover, increased landlessness need not necessarily lead to increased incidence of poverty. If aggregate demand for labour increases sufficiently rapidly and agriculture grows fast enough, real wages and earnings of the landless may continue to rise. Finally, it should be noted that an increase in land per person is no guarantee against increased landlessness and poverty. Increase in landlessness and poverty can occur, despite an improvement in the overall land–person ratio, if changes in the distribution of access to land are sufficiently adverse.

This chapter includes discussion of the effect of demographic growth in Asia on (*a*) the distribution of access to land and (*b*) rural poverty. Differences in the experiences of the major countries are highlighted, and the 'escape routes' identified for the typical case in the countries that succeeded in avoiding a part or the whole of the aforementioned path. The escape routes are further examined to see whether they are merely avenues for temporary adjustment to a high rate of demographic expansion or whether they are permanent mechanisms to cope indefinitely with such expansion. Finally, the case of a country that essentially represents the preceding archetypical case is explored in some detail. In conclusion, the chapter summarizes the implications of the varied experiences of Asian countries for policies to ensure equitable access to land.

The Experience in Selected Asian Countries

The experiences of eight Asian countries are explored: Bangladesh, China, India, South Korea (referred to as Korea in the rest of the chapter), Pakistan, Philippines, Taiwan, and Thailand. To the extent that the availability of data permits us to distinguish between the dissimilar experiences of Java and the rest of the islands, the experience of Indonesia is also mentioned occa-

Table 7.1   Population growth in selected Asian countries (annual compound rate of growth)

Country	1950–60	1960–70	1970–81
Bangladesh	1.9	2.5	2.6
China	1.8	2.3	1.7
India	1.9	2.3	2.1
Korea	2.1	2.6	1.7
Pakistan	2.3	2.8	3.0
The Philippines	3.0	3.0	2.7
Taiwan	3.5	3.1	2.0
Thailand	2.7	3.0	2.5
Indonesia	2.1	2.1	2.3

sionally. Together, the experiences of these countries encompass a great variety of interaction between demographic and other factors, producing a number of distinctly different outcomes.[2]

In all of these countries except Taiwan, population growth rate accelerated between the 1950s and the 1960s (see Table 7.1).[3] In several countries (Bangladesh, Pakistan, and Indonesia), it continued to accelerate in the 1970s. Only in China and Korea (and, to a lesser extent, in Taiwan) does the growth rate appear to have moderated somewhat in the 1970s. Everywhere else, the rate of growth continues to be above 2 per cent per year. As Kuznets noted long ago, the rates of population growth in the industrialized countries during their development periods were far lower (Kuznets 1956).

Because population growth (with a given time-lag) translates into growth in labour force, the rate of growth in the labour force in most of these countries will continue to accelerate in the near future and will continue to be high until at least the turn of the century, regardless of the success in reducing population growth in future. Since the 1950s, the labour force consistently increased at a faster rate than did population.[4] However, the uniformity among the countries ends here; the chain of effects of demographic expansion on the land–person ratio, landlessness, and standards of living varies a good deal among them. A summary of the main trends in each country follows (see Table 7.2 for data on each country).

Bangladesh

Bangladesh strongly resembles the archetypical case described in the introduction. Rapid population growth led to a high rate of increase in

---

[2] Of all the large countries of the region—that is, those with more than 20 million population in the early 1980s—only Burma and Vietnam are excluded.

[3] In Indonesia, it remained the same in the 1960s as it was in the 1950s, but it accelerated in the 1970s. Taiwan's growth rate in the 1960s, though lower than in the 1950s, was higher than 3 per cent per year.

[4] For example, the labour force in China increased at annual rates of 3.3 per cent in the 1950s, 2.9 per cent in the 1960s, and 2.1 per cent between 1970 and 1981.

Table 7.2   Basic country data

Country	Year	Net sown area (thousand ha)	Gross cropped area (thousand ha)	Agricultural labourers (thousands)	Net sown area per labourer (ha)	Gross cropped area per labourer (ha)
Bangladesh	1960/1	8,437	11,100	14,523	0.58	0.76
	1980/1	8,563	13,161	19,906	0.43	0.66
China	1952	107,829	141,256	173,170	0.62	0.82
	1960	104,040	150,575	170,190	0.69	0.88
	1980	97,200	146,379	302,110	0.32	0.48
India	1950/1	118,800	133,200	97,200	1.22	1.37
	1960/1	133,500	152,800	131,100	1.02	1.17
	1980/1	142,300	174,000	146,800	0.97	1.19
Pakistan	1960/1	12,995	13,975	8,945	1.45	1.56
	1978/9	15,330	19,160	12,197	1.26	1.57

labour force. Industrialization was not sufficiently high or labour-absorbing to prevent a rapid increase in the labour force in agriculture. Total supply of cultivated land increased very little. As a result, the land–labour ratio in agriculture declined. Between 1960 and 1980, it is estimated to have declined by one-quarter.[5] The same is true (though to a lesser extent) of the change in cropped area of land per agricultural labourer. In the absence of any serious redistributive institutional measure, landlessness increased as the poorer landowners were forced to sell land, a process that accelerated during periodic famine and near-famine years. An estimate by the present author shows that it increased, as a percentage of total farming labour force, from 14 per cent in the early 1950s to 26 per cent in the late 1970s (Ghai, Khan, Lee, and Radwan 1979). Within agriculture, demand for labour increased at a very slow rate: slow growth in output was one factor responsible for it, and also, employment per unit of land remained lower than would be feasible and desirable due to the prevalence of inappropriate institutions and the absence of appropriate technological advances.[6] Non-agricultural employment in rural areas did not increase much. As a result, the gap between the supply of and the demand for labour in agriculture increased, and real wages fell. Between the first half of the 1950s and the second half of the 1970s, real wages in agriculture fell by one-quarter (Khan and Lee 1984, 185–203), and there is little evidence of any offsetting factor (such as increased employment per person or per family of a given size). The incidence of *poverty* (defined as

[5] This appears to be an underestimate. It has been argued by many, including the present author, that the method of estimating the agricultural labour force in the 1970s amounted to an underestimation, relative to that in earlier years.

[6] Some discussion of such factors is to be found in ARTEP (1980) and Khan and Lee (1981).

the proportion of population belonging to households with incomes below the amounts needed to provide basic nutrition and related consumption) increased over the years.

China

Cultivated land per member of agricultural labour force is about the lowest in China. Land–person ratio fell most dramatically between 1960 and 1980, cultivated land per agricultural worker by 48 per cent and cropped land per agricultural worker by 41 per cent. This was due to the declining supply of cultivated and cropped land and the failure to transfer labour out of agriculture and into industrial and modern tertiary sectors. At least on the surface, the situation appears to have been better in the 1950s. During the first decade after liberation, cropped land increased. Also, with the onset of the Great Leap Forward, there was a sharp temporary rise (fall) in the proportion of the labour force employed in industries (agriculture). This partially continued until 1960 (see Table 7.3). Between 1960 and 1980, the agricultural sector's share of the labour force increased from 66 per cent to 72 per cent. Cultivated land and cropped land fell by 7 per cent and 3 per cent respectively.

Table 7.3   Change in the share of the industrial sector in the labour force

Year	Per cent	Year	Per cent	Year	Per cent
1952	6.0	1959	11.0	1962	6.6
1957	5.9	1960	11.5	1970	8.2
1958	16.6	1961	8.7	1980	13.4

Note: Changes between 1970 and 1980 and since 1980 have been smooth.

However, the equality of access of land of those dependent on agriculture was maintained: first, by the institution of collective agriculture under the commune system, and, since the replacement of communal agriculture by a system of contracting with individual households, by adhering to the principle of equal amount of land per capita and/or per worker.[7] This prevented an increase in the incidence of poverty. Indeed, there was a dramatic reduction in the incidence of absolute poverty in the early years and, with the exception of the few years of difficulty in the late 1950s and early 1960s, this achievement was maintained and improved upon in later years.

[7] The individual contracting system that came into existence in 1979 now covers more than 95 per cent of rural households. The system resembles fixed rental tenancy. For an account, see Khan and Lee 1983. It should be noted that equality of access to land was ensured only in local communities, both under the commune system and under the current contracting system. It was almost perfectly equal within the so-called basic accounting unit (usually a team consisting of fewer than 30 households). Between areas of differing land–person ratio, there clearly were substantial inequalities of access.

The sharp reduction in land–labour ratio meant a corresponding increase in labour input in agriculture. While output per worker (that is, per person-year) appears to have increased slightly, output per person-day appears to have declined quite sharply between the mid-1950s and the mid-1970s.[8] Since the institution of the system of contracting, there has been substantial improvement in the value of output per worker due to the combined effect of increased production promoted by the systemic change and the sharp improvement in the terms of trade that accompanied such change. However, continued growth of agriculture in the future does not appear to be assured without a technological transformation. The crucial element in such transformation is the reversal of the declining trend in the land–labour ratio by increasing non-agricultural rural employment and, most importantly, reducing the growth of the labour force through population control.

India

Initial land endowment in India was far more favourable than in the afore-mentioned countries and in Asia as a whole. In the three decades between 1950/1 and 1980/1, net cultivated area per member of the agricultural labour force declined by 20 per cent. Much of this decline took place in the 1950s; the decline since 1960/1 has been very modest. Gross cropped area per agricultural labourer declined by about 15 per cent during the 1950s and changed little after 1960/1. The supply of land (gross cropped area) increased at a modest rate of 0.8 per cent per year over the three decades, although the rate of increase slowed down to 0.4 per cent per year during the 1970s. Since 1960/1, the growth of the labour force in agriculture has been slow. The combined effect of these factors has been to reduce net cultivated land per worker modestly between 1960/1 and 1980/1 and to leave gross cropped land per worker unchanged.[9]

It would appear that the pure effect of demographic pressure on land was not great enough to cause a large increase in landlessness—yet it appears that the landlessness increased very substantially. Reliable and intertemporally comparable time series for a sufficiently long period are not available. However, between 1964/5 and 1974/5 the number of agricultural labour households without land increased by 23 per cent. During the same period, the

---

[8] See Rawski 1979, 120. Rawski estimates that between 1957 and 1975 the gross value of agricultural output per person-year increased by 10 per cent and that the value per person-day declined by 15 per cent to 36 per cent (depending on assumptions made about the intensity of labour use in certain cropping operations).

[9] One should note that, compared to 1960/1, the labour force in 1980/1 might have been seriously underestimated. Between these years, the labour force participation rate declined from 43 per cent to 33.4 per cent. It is doubtful whether this reported decline reflects reality. Thus, land–labour ratio may have declined at a somewhat greater rate than is revealed by the estimates based on the reported data.

number of agricultural *labourers* (people belonging to households with little or no land) increased by 49 per cent (The Government of India 1975).

The effect on the incidence of poverty varied from one region to another, depending on the performance of the agricultural sector. Two extreme examples are the Punjab and Bihar. In the Punjab, the trend in access to land appears vastly dissimilar, depending on whether the ownership or the operational holding is considered the criterion and on whether complete landlessness (meaning no access to land) or near-landlessness (meaning access to less than one acre per family) is used as the measure. Between 1953/4 and 1971/2, the proportion of households owning *no* land declined from 37 per cent of the total to 9 per cent; but the proportion of households owning *less than one acre* increased from 14 to 49 per cent. During the same period, the proportion of households having no operational holding increased from 29 to 54 per cent, but the proportion of households having less than one acre of operational holding declined from 16 to 2 per cent.[10] The combined proportion of landless and nearly landless households—on either criterion—increased: according to the ownership criterion, from 51 to 58 per cent, and according to the operational holding criterion, from 45 to 56 per cent. The incidence of rural poverty appears to have declined during the 1960s and 1970s due to the high rate of growth of agricultural output, the increased demand for labour both in agriculture and in rural non-agricultural activities, and the consequent rise in real wages and earnings.

In Bihar, the effect of demographic expansion on the distribution of access to land has been quite dramatic. Between 1964/5 and 1974/5, the rate of increase was 16 per cent for all rural households and 33 per cent for rural labour households (that is, those having little or no land). Male agricultural labourers increased by 91 per cent between 1961 and 1971. Real wages declined. The incidence of rural poverty remained undiminished at a very high level, or it even increased.

Pakistan

Since the 1950s, the population growth rate in Pakistan has been one of the highest in the region. Since the late 1970s it has had a rate of growth higher than that of any other country in the region. During each successive decade, the population growth rate accelerated, and this translated itself into a high rate of growth of the labour force. The labour force in agriculture, however, grew at a relatively modest rate of 1.74 per cent per year between 1960/1 and 1978/9, due to a reasonably rapid absorption of labour in the non-agricultural sectors and, especially since the mid-1970s, a high rate of emigration

---

[10] These changes were probably partly the effect of land reform legislations in the Punjab. For a further account, see Sudipto Mundle 1984.

to the Middle Eastern countries. Also, the supply of agricultural land increased due to the expansion of both net sown area and cropping intensity. Between 1960/1 and 1978/9, net sown area per agricultural worker declined by 13 per cent, but gross cropped area per agricultural worker remained unchanged.

Pure demographic pressure for increased proletarianization of the peasantry would appear to have been modest. And yet it appears that during the 1960s and 1970s large-scale proletarianization took place in Pakistan. Direct estimates are not available. Naseem's indirect estimate of a 357 per cent increase in landless labour between 1960/1 and 1972 is generally regarded to be too high. However, the more cautious estimate by Hussain claims that 43 per cent of the total number of agricultural labourers in 1973 had entered this category as a result of the pressures toward proletarianization of the peasantry over the preceding decade. Indirect estimates for more recent periods suggest that the trend toward increased landlessness continued in the 1980s.[11] The main pressure toward increased proletarianization came from the changing agrarian structure in the wake of the Green Revolution. Higher profits led to large-scale eviction of tenant farmers, who joined the ranks of wage labourers. Also, the prosperity of the Green Revolution was distributed highly asymmetrically among the beneficiaries, making it necessary for many small farmers to supplement agricultural income with part-time wage labour.[12]

Increased landlessness does not appear to have depressed real wages. Indeed, real wages increased during much of the 1960s and 1970s. The explanation lies in the increased demand for labour, due to the increased labour requirement of the Green Revolution technology in the 1960s (tractors being very limited before the 1970s) and large-scale emigration in the 1970s. Thus, increased proletarianization of the peasantry coincided with the declining incidence of absolute poverty among the rural population.

Thailand

Thailand also experienced a very high rate of population growth (see Table 7.4). Industrialization succeeded in reducing somewhat the share of agriculture in the total labour force. Yet the annual rate of increase of agricultural labour was close to 2 per cent between the years 1960 and 1975. However, the most outstanding feature of Thailand's rural development is the phenomenal expansion in the area of farm land. Between 1960 and 1975, total farm land increased at an annual rate of 4.3 per cent. Farm land per agricultural worker—already high by Asian standards—increased by 41 per cent over the period under consideration.

[11] See Irfan and Amjad 1984 for a discussion of the various sources of evidence, including the estimates made by Naseem and Hussain.
[12] For an analysis of these changes, see Irfan and Amjad 1984.

Table 7.4   Thailand data

Year	Total farm land (thousand ha)	Rural population (thousands)	Workers in agriculture (thousands)	Farm land per rural person (ha)	Farm land per agricultural worker (ha)
1960	9,869	22,970	11,332	0.43	0.87
1970	14,853	31,300	14,192	0.48	1.05
1975	18,605	36,010	15,132	0.51	1.23

Initial agrarian structure was characterized by the low incidence of land-lessness. Rural institutions did not experience any major unfavourable change. Direct estimates of the incidence of landlessness are not available, but it is widely believed to be around 10 per cent of the rural households. Together with the fact that agriculture grew rapidly, the preceding factors explain the significant reduction in the incidence of rural poverty in Thailand since the 1960s.

The Philippines

The trend in the land–labour ratio and the factors behind it in the Philippines are similar—though somewhat less strongly so—to those in Thailand (see Table 7.5). Rapid demographic expansion translated itself into a somewhat slower—though very high by any absolute standard—rate of increase in the agricultural labour force. The annual rate of increase in the period between 1956 and 1977 was 2.1 per cent. However, the increase in the amount of land was significantly higher: 2.7 per cent per year over the same period. The planted land per person employed in agriculture increased by 12 per cent between 1956 and 1977.

Table 7.5   Data on the Philippines

Year	Planted land (thousand ha)	Employment in agriculture, forestry, hunting, and fishing (thousands)	Land per worker (ha)
1956	6,814	4,548	1.50
1960	7,594	5,224	1.45
1970	8,946	6,212	1.44
1977	11,843	7,046	1.68

The similarity to the experience of Thailand ends at this point. Not enough data exist to enable us to document the level and/or trend in landlessness or the incidence of rural poverty. But enough scattered evidence exists to suggest that landlessness has been increasing, real wages have been falling, and

Table 7.6   Data on Taiwan

Year	Cultivated area (thousand ha)	Gross cropped area (thousands ha)	Farm population (thousands)	Agricultural workers (thousands)	Land per capita of farm population		Land per agricultural worker	
					Cultivated (ha)	Cropped (ha)	Cultivated (ha)	Cropped (ha)
1946	832	975	3,522	1,285	0.24	0.28	0.65	0.76
1950	871	1,435	3,998	1,414	0.22	0.36	0.62	1.01
1955	873	1,508	4,603	1,489	0.19	0.33	0.59	1.01
1960	869	1,600	5,373	1,464	0.16	0.30	0.59	1.09
1970	905	1,656	5,997	1,520	0.15	0.28	0.60	1.09
1975	917	1,659	5,598	1,487	0.16	0.30	0.62	1.12

the incidence of rural poverty has become more widespread.[13] Clearly, these were not due to demographic pressures on land. The explanation lies in the highly unequal distribution of land ownership, the high initial landlessness, and the institutional and technological obstacles to increased labour use in agriculture.

Taiwan

Taiwan's population grew at extraordinarily high rates during the 1950s and the 1960s (see Table 7.6). Supply of cultivated land has been virtually static. There was a jump in cropping intensity in the late 1940s; thereafter, cropped area has also increased at a very slow rate. Since 1960, the annual increase in cropped area has been less than 0.25 per cent. The amount of cultivated land per unit of agricultural population declined steadily and the amount of cropped land per unit also declined over much of the time period. However, cultivated land per agricultural worker remained pretty much unchanged. Cropped land per agricultural worker increased moderately since the late 1950s, due to the rapid industrialization and the increase in rural non-agricultural output and employment. The latter factor deserves particular attention: for more recent dates, its importance is demonstrated by the declining share of agricultural income in total farm family income, from 66 per cent in 1966 to 27 per cent in 1979 (Kuo 1983).

Several other features of the Taiwan experience deserve special attention. In the early 1960s, demographic pressure on land was building up. The implementation of a highly egalitarian land reform at that time must have had a crucial role in ensuring access to land during that transitional phase. Second, the extreme initial scarcity of land was eased substantially by the high—and highly productive—absorption of labour into Taiwanese agriculture.

Korea

The Korean experience underlines the crucial importance of rapid industrialization. In the 1950s, the land–person and land–labour ratios in Korea's agriculture were the least favourable of all Asian countries, although probably mitigated to some extent by the high absorption of labour per unit of land (see Table 7.7). Land supply was virtually static over time, while the population increased rapidly in the 1950s and 1960s. But, except in the earlier years, overall demographic growth did not lead to an increased pressure

---

[13] For evidence up to the early 1970s, see ILO 1977, chapter 9. Official wage data indicate a continuing decline in real wages. Direct estimates of landlessness relate to local areas. Thus, for example, an IRRI study in a *barrio* (neighbourhood) in Laguna shows that landless worker households, as a proportion of all households, increased from 30 per cent in 1966 to 50 per cent in 1976. See Publico 1978.

Table 7.7   Data on Korea

Year	Cultivated area (thousand ha)	Farm population (thousands)	Agricultural employment (thousands)	Land	
				Per farm population (ha)	Per agricultural worker (ha)
1955	1,994	13,300	—	0.15	—
1960	2,025	14,559	6,775	0.14	0.30
1980	2,196	10,827	4,658	0.20	0.47

on land. Since 1960, there have been large absolute reductions in both the farm population and agricultural employment. Land–person and land–labour ratios in agriculture increased very substantially between 1960 and 1980. The other important fact to note in the Korean experience is the role that egalitarian land reforms played in the early days in ensuring access to land at a time when demographic pressures on land had been building up.

Summary

The preceding eight cases present a wide variety of experiences. Most other cases in South and East Asia can be described with reference to one of these experiences or a combination of several. In Indonesia, for example, the experience of densely populated Java has been quite different from that of the other islands, where there is a relative abundance of land. The extreme land scarcity of Java is comparable with that of Bangladesh. Its experience of expanding population, leading to a continuous decline in the already low land–person ratio, led social scientists to describe it in terms of the models of 'agricultural involution' and 'shared poverty' (Geertz 1963). Land supply has remained static for a long time. The continued high rate of population growth was translated into an annual rate of growth of the rural population totalling slightly less than 2 per cent since the mid-1960s. The cultivated area per agricultural worker declined from 0.42 ha in 1963 to 0.33 in 1973. The degree of landlessness has been very high, estimates ranging up to 50 per cent of rural households. Compared to the rest of contemporary developing Asia, labour use per hectare of land and output of rice per hectare have been high. Nevertheless, wages have been low and poverty widespread. In the absence of clear statistical evidence, considerable controversy continues on the trend in rural poverty since the mid-1960s. Some have argued that growth elsewhere in the economy and in agriculture itself provided high employment and contributed to the alleviation of poverty. Others argue—somewhat more convincingly—that available evidence does not indicate any such outcome. The effects of the declining land–labour ratio and the high incidence of landlessness were exacerbated by unfavourable changes in technology (such as that in harvesting rice) and institution (like the abolition of the traditional right of labourers to participate in paddy harvest and receive a proportion of it as earnings), which have led to lower employment and a worse distribution of income.

The Longer-term Perspective: The Bangladesh Case

In the preceding discussion, it was clearly demonstrated that Bangladesh is the most extreme example of demographic pressure leading to reduced land–person ratio, declining access to land, increased landlessness, and higher incidence of absolute poverty. In the list of countries, it was the only one in which the linkage at every stage conformed to the archetypical scenario.[14] It is, therefore, of some importance and interest to take a closer look at the Bangladesh experience in a somewhat longer historical perspective.[15]

In Table 7.8, an attempt has been made to show the trend in land–person ratio in Bangladesh over the last ninety years. Land–labour ratio in agriculture must have moved very closely with this index. It was only since the 1960s that its rate of decline was a shade slower than that in the land–person ratio. The decline in the land–person ratio has been a long-term phenomenon, but the rate of decline was modest until 1951. Since 1951, the rate of decline accelerated rapidly due to a sharp acceleration in population growth. In the thirty years since 1951, the decline has been 51 per cent.[16]

In the nineteenth century, pure landless labourers were an insignificant proportion of the agricultural labour force. Agricultural labourers (that is, wage-earners) were largely those who were engaged in farming on small

Table 7.8   Land–person ratio in Bangladesh

Year	Index of population	Index of net sown area	Land–person ratio index
1890	63	87	138
1901	69	89	129
1931	85	96	113
1941	100	98	98
1951	100	100	100
1981	208	102	49

[14] The same result appears to have obtained in parts of India. The experience of Nepal in terms of poverty and access to land may have been similar, although the land–labour ratio does not seen to have gone down in recent years. The initial condition in Java was in many ways similar to that in Bangladesh. Many of the trends were similar, although the final outcome in terms of trends in poverty is much less extreme, if not altogether different.

[15] This part of the chapter is drawn largely from the author's chapter on the real wages of agricultural workers in Bangladesh (Khan and Lee 1984, chapter 8).

[16] Land has been measured in terms of net sown area. If it were measured in gross cropped area (allowing for the increase due to the rise in the intensity of cropping), the decline would be less sharp, though still accelerating at about the same time. For earlier years, estimates of gross cropped area are not available.

owned or tenanted farms. Wage labour was their additional occupation, perhaps the principal one for those who had tiny farms.[17] By the early 1930s, concern began to be expressed about the sharp increase in the number of landless agricultural labourers in Bengal. This was based on the observation that the number of agricultural workers in Bengal increased by 60 per cent between the population census of 1921 and that of 1931.[18] The present author has estimated elsewhere that landless labourers as a percentage of total farming labour force have increased steadily since 1951 and that the rate of increase has accelerated in more recent years. It increased from 14.3 per cent in 1951 to 19.8 in 1967–8 and 26.1 per cent in 1977 (Ghai *et al.* 1979, 149).

The increased landlessness must be seen as the direct result of the demographic pressure on limited and non-expanding land in the absence of any redistributive institutional change. There is little evidence of a significant reduction in the inequality of access to land over the decades.

The course of real wages in agriculture and the incidence of rural poverty have been documented reasonably comprehensively since the mid-1950s. Considerable fluctuation in the magnitudes of these indicators occurred. However, the long-term trend is one of clear deterioration. Real wages in the early 1980s were about 25 per cent below those in the early 1950s. Absolute poverty increased correspondingly.

Rudimentary measurements show that the decline in real wages observed since 1950 was probably preceded by a very long period of declining real wages in Bangladesh agriculture. In terms of cleaned rice equivalent, real wage in the 1830s was about 6 kg. In the late 1880s, it ranged between 4.2 and 5.2 kg. Even in the late 1930s, it was approximately 6 kg (Khan and Lee 1984, 197–200; cites all sources). Around 1950, it was about 3 kg, and in the 1970s it fell below 2.5 kg. Without claiming any knowledge of the complete course of the movement in real wages over the century and a half, one can surmise that, for substantial periods in the last century and the earlier part of the present century, real wages were as high as is indicated by the range of estimates shown earlier, and that no record is available to show, at any time between

[17] This has been repeatedly suggested in an 1888 Government of Bengal report. For example, the commissioner of the Chittagong division wrote: 'The class of agricultural laborer as distinguished from the cultivator may be said not to exist here; though a very large number of cultivators also work as agricultural laborers'. The settlement officer of Nulchira wrote, 'There are no landless day laborers in Nulchira . . . every man holds or possesses a share in a piece of land sufficient to prevent his being compelled to work as a day laborer for daily wages.'

[18] See M. Azizul Huque 1980 (originally published 1939), 132–4. During this period, the rate of decline in the land–person ratio was still modest. One wonders if a part of the reported increase in agricultural labourers was not due to such spurious factors as changed definition and changed response on the part of the landowners/landlords, who tended more to classify tenants as labourers in 1931 as compared to 1921. It is well known that such factors have been in operation in contemporary India. In Bengal, such a response may have been evoked by the tenancy legislation that was attempted circa 1928 and that may well have appeared as a threat to the landowners *vis-à-vis* the tenants.

the 1830s and the 1930, that real wages, in rice equivalent, were as low as the observed levels since 1950. It also seems unlikely that a significant part of the long-term decline in the preceding measure of real wages can be explained with reference to a rise in the relative price of rice. Furthermore, it appears to be reasonably clear that the decline in real wages, both over the longer term and over the more carefully documented period of the past thirty years, signifies an increased impoverishment of the rural population. Over time, the proportion of rural households dependent on wages as a main source of income has been increasing, and there is no indication that employment per consumption unit has increased.

Declining land–person ratio must be seen as the major factor behind this outcome. Its effects were exacerbated by the continued inequality in the distribution of land,[19] the prevalence of institutions and techniques discouraging the absorption of labour in agriculture, the slow growth of agricultural output, and the failure of non-agricultural employment (both in the rural and the urban areas) to expand.

What may appear baffling is that the sustained impoverishment of a large and increasing section of the population has been consistent with fairly rapid demographic expansion. In particular, since the mid-1950s, the rate of population growth has been phenomenal—a compound rate of nearly 2.5 per cent per year—in spite of a falling consumption standard for a very large proportion of the population.[20] According to available evidence, only a part of the increased growth in population was due to the declining death rate. Thus, in 1931, Bengal is reported to have experienced a birth rate of 27.8 per 1,000 and a death rate of 22.3 per 1,000 (The Government of India 1931; quoted in Huque 1980, 184). These compare with an average birth rate of about 42 and death rate of 17 since the 1960s.[21] Some of the decline in the death rate is perhaps accounted for by the expansion of public health

[19] In 1950 a land reform was enacted in Bangladesh (then East Pakistan). It abolished the *zaminders* (landlords who collected revenue from peasant tenants) and conferred land ownership on the tenants. This, however, did not result in an equitable distribution of land ownership, as the distribution of land-holding among the former tenants had been very unequal. This was in sharp contrast to the East Asian case as exemplified by South Korea where, in the pre-land-reform period, the distribution of land-holding among the actual operators was highly equal, though that among the owners was not. Land reform, granting ownership to the actual operators, resulted in a highly equal distribution of land ownership in South Korea.

[20] Apart from the falling wage rate, a good deal of information, albeit fragmentary and rudimentary, suggests that the typical level of food consumption today is less than what it was a century or half a century ago. Today, the customary per capita food-grain consumption figure used in making the food plan of the government is between 15 and 15½ oz. per day. In all the aforementioned historical documents and reports, the norms used for reference have been way above this figure. The few family consumption figures available in the aforementioned 1888 report also indicate a much higher per capita consumption. Finally, one should refer to the Bengal Famine Code (revised edition, 1905), which calculated wages and the famine ration on the basis of a 33 oz. (16 *chhataks*) per capita daily requirement of rice for an adult worker, a figure that would imply a substantially higher average requirement norm per person than the ones in current use.

[21] The religious/cultural differences between the population of Bengal in the 1930s and that of Bangladesh since the 1950s could explain a part of the increase in the birth rate, but only a part.

measures, such as the increased supply of drinking water and the (largely consequent) eradication of epidemics. The observed sharp reduction in mortality would, however, be hard to explain with reference to the expansion of these measures if large groups of the population experienced a decline in nutritional standards proportionate to the long-term decline in real wages. It is probable that several types of responses for survival were activated to reduce the decline in nutrition to a rate slower than that in real wages.

A classic survival mechanism would be an increase in the participation rate and the number of days of employment. Conventional measurements of these—especially the wage-based components—are so low at the present time that it is hard to conceive that over time there has been much rise in them. In more recent years, the participation rate (the proportion of population in the employed labour force) seems to have fallen from 34 per cent in 1961 to 28 per cent in 1974. The latter is one of the lowest recorded rates in contemporary South and South-east Asia.

It is probable, however, that there has been an increase in subsistence activities that are not caught either by labour force surveys or by population censuses as either economic activity or an employment category. These activities have probably taken the form of the gathering of inferior food and fuel and self-employment of a very rudimentary type for subsistence production. As a result of all these activities, some augmentation of nutrition (through the collection of inferior vegetables, tubers, and fruits) and of income may have taken place.

Another factor contributing to a smaller reduction in nutrition (particularly in energy intake) than in real wages was perhaps a decline in the perceived quality of diet.[22] One element of this is the rise in the share of wheat in the diet of the poor. In recent years, there has been a sharper than average decline in the quantities of meat, fish, fats, and oils.[23] It also seems likely that the real consumption of non-food items declined more sharply than food.

All these factors must have contributed to some protection of energy intake in spite of declining real wages. It is clear, however, that, as far as the rural poor are concerned, a substantial decline in energy intake has taken place since the 1950s. If the evidence of the longer-term decline in real wages is to be believed, then such a decline may have been going on over a very long period of time.

---

[22] This is a reduction in quality not from a dietary point of view. For example, wheat has slightly fewer calories but substantially more protein per unit of weight as compared to rice. The increased share of wheat in total cereals is seen as an indicator of declining quality from the standpoint of consumer preference and cost.

[23] This is based on the comparison of consumption of income groups corresponding to the wage-earners, as reported by the household expenditure surveys of 1963/4 and 1973/4.

Some Conclusions

The Asian experience shows considerable variety in the effect that demographic expansion had on access to land and the livelihood of population. The Bangladesh experience clearly represents the archetypical case: demographic expansion successively leading to a reduction in the land–person ratio, a reduced access to land, increased proletarianization of the peasantry, and an increased incidence of poverty. However, actual experience in most countries varied a good deal from this archetypical case. These facts, and their implications for policy, are summarized in the following short propositions.

One obvious escape route is the expansion in land. This was the main factor behind the ability of Thailand and the Philippines to avoid a decline in the land–person and land–labour ratios.

Another way to escape the effect of demographic expansion is rapid industrialization, which leads to stable or declining numbers employed in agriculture and/or dependent on agriculture. This was most convincingly demonstrated by Korea and Taiwan.

A related avenue is the expansion of non-agricultural employment in the rural areas. This was an important element of the transition in Taiwan. It was also featured in China, partially to mitigate the effect of the crushing decline in land–labour ratio. It may also have played an important role in making the outcome in Java so very different from that in Bangladesh.

Another possibility is to increase the productive absorption of labour into agriculture itself. This possibility is indicated by the fact that labour input per hectare of land in contemporary developing Asia is substantially lower than what it used to be in some East Asian countries (notably Japan) at a comparable stage of development and that at best a part of the difference can be explained by agro-climatic factors.[24] This was an important factor in Korea and Taiwan in mitigating the extremely low initial land–labour ratio. In Java, high labour absorption per hectare has, to a certain extent, blunted the effect of the low land–person ratio. In China, this factor had a major role in providing employment during a period of rapidly declining land–labour ratio.

Redistributive institutional measures can successfully preserve, or even expand, access to land in the face of the declining land–person ratio due to increasing demographic pressure. China is a classic example of this. We have also argued that in both Korea and Taiwan egalitarian land reform had a critically important role in ensuring equitable and widespread access to land during the early years of adjustment to high demographic pressure.

Reduced access to land need not necessarily translate itself into increased misery and poverty. Increased demand for labour may still lead to an

[24] Literature on this is quite considerable (references in Khan and Lee 1981).

improvement in real wages and earnings. Recent experience in Pakistan (and possibly in the Indian Punjab) can be cited as an example.

An important question is whether the preceding six offsetting factors are alternatives to a reduction in population growth in any realistic sense in the long run. Even in Thailand and the Philippines, the expansion in land is believed to be coming to an end. Elsewhere in Asia, this has been a very limited possibility and usually an expensive one. Only an extraordinarily rapid rate of industrialization—as exemplified by the unprecedented experience of Korea and Taiwan—can effect the net transfer of labour out of agriculture. Even these countries felt the need to reduce the rate of population growth, and Taiwan at best succeeded in avoiding an increase in agricultural employment. Elsewhere in Asia, fairly rapid rates of industrial growth failed to relieve agriculture of demographic pressure. China is an outstanding example. China is also an example of the limit of productive absorption of labour into agriculture: as has already been seen, output per unit of labour input declined as a result of a massive increase in the latter. Finally, redistributive and institutional measures also have their limits in a situation of declining land–labour ratio. Once again, China illustrates the point. Despite its remarkable achievements, collective agriculture could not continue to provide the vastly increased population that came to be dependent on it with a continuously improving living standard at an acceptable rate.[25] It is in this context that the extreme and unorthodox Chinese policies to reduce the growth rate of population are best understood.

It is clear, therefore, that the Asian countries must have the policy of reducing the growth rate of population as the central component of the strategy of ensuring continued access to land. A reduction in population growth, however, will have an effect on the growth of the labour force only after a long time-lag. The growth rate of the labour force until the turn of the century and beyond has already been determined in these countries. Even if these countries are able to bring down the rate of population growth to a viable long-term level in the next five to ten years, they will have to cope with high rates of growth of the labour force for the next twenty-five years. To make this adjustment, they will need to seek solutions along the aforementioned 'escape routes'. During this transition period, emphasis has to be placed on increased absorption of labour into agriculture, increased employment in rural non-agricultural activities, adequate labour intensity of indus-

---

[25] Often one encounters the argument that, in a densely populated agrarian society, collectivization is the only feasible institution for the insurance of egalitarian access to land, as there is not enough land to provide everyone with the minimum required size of plot under peasant farming. As long as the same amount of labour must be absorbed on the same area of land, it is hard to see how collectivization by itself can solve the problem. It is difficult to argue, either a priori or on the basis of actual evidence, that in agricultural operations collectives ensure greater and more effective labour absorption than does peasant farming. In capital construction, collectives have the advantage in mobilizing more labour; but, in practice, it has not been found to be easy without running into problems of organization of incentives.

trial projects, and redistributive institutional reforms to ensure reasonably equitable and wide access to land in a situation of growing land scarcity.

The importance of redistributive institutional reforms is properly gauged once it is recognized that the successful avoidance of demographic pressure on land and/or the improvement in the land–person ratio does not guarantee equitable access to land. The increased proletarianization in the Philippines cannot be attributed to demographic pressure on land. In Pakistan, the increased proletarianization of the peasantry was at best partly due to demographic pressure. In large part, it was due to an institutional transformation of agriculture. Adequately large and effective institutional reforms are needed to counter these forces.

References

ARTEP (1980), *Employment Expansion in Asian Agriculture: A Comparative Analysis of South Asian Countries*, Bangkok.

Geertz, C. (1963), *Agricultural Involution: The Process of Ecological Change in Indonesia*, Berkeley, Calif.

Ghai, D. P., A. R. Khan, E. Lee, and S. Radwan (eds.), (1979), *Agrarian Systems and Rural Development*, Macmillan.

The Government of Bengal (1888), *Report on the Condition of the Lower Classes of Population in Bengal*, Calcutta.

The Government of India (1931), *Report of the Public Health Commissioner*.

—— (1975), *Rural Labor Enquiry 1974–75: Final Report on Wages and Earnings of Rural Labor Households*.

Huque, M. A. (1980), *The Man behind the Plough*, Dhaka.

ILO (1977), *Poverty and Landlessness in Rural Asia*, Geneva.

Irfan, M., and R. Amjad (1984), in A. R. Khan and E. Lee (eds.), *Poverty in Rural Asia*, Bangkok.

Ishikawa, S. (1967), *Economic Development in Asian Perspective*, Tokyo.

Khan, A. R., and E. Lee (1981), *The Expansion of Productive Employment in Agriculture: The Relevance of the East Asian Experience for Developing Asian Countries*, ILO/ARTEP, Bangkok.

—— (1983), *Agrarian Policies and Institutions in China after Mao*, ILO/ARTEP, Bangkok.

—— (eds.) (1984), *Poverty in Rural Asia*, Bangkok.

Kuo, S. W. Y. (1983), *The Taiwan Economy in Transition*, Boulder, Colo.

Kuznets, S. (1956), 'Quantitative Aspects of Economic Growth of Nations', *Economic Development and Cultural Change*, October.

Mundle, S. (1984), 'Land, Labor and the Level of Living in Rural Punjab', in A. R. Khan and E. Lee (eds.), *Poverty in Rural Asia*, Bangkok.

Publico, S. M. (1978), 'Landless Agricultural Laborers Left out in the Cold?', *Gintong Butil* (NGA monthly publication), October.

Rawski, T. G. (1979), *Economic Growth and Employment in China*, New York.

8 The Impact of Demographic Changes on Rural Development in Malawi

GRAHAM CHIPANDE

Economics Department, Chancellor College, University of Malawi, Zomba

It is widely agreed that the growth strategy pursued in most of the developing countries during the 1950s and 1960s resulted in a development pattern whereby most of the rural masses did not share in the benefits of the development effort (see Lipton 1977 and World Bank 1975*a*). The 1970s were therefore the decade during which many developing countries (including Malawi) launched rural development programmes, mostly with the help of the World Bank and other agencies, in an attempt to bring the benefits of economic development to the rural poor. Because agriculture is the predominant activity among rural populations in most developing countries, agricultural development has become a *sine qua non* of rural development (Massignon 1973). Smallholder agricultural development has been the most common strategy undertaken by many developing countries, especially in Sub-Saharan Africa, to improve rural lives (although such aspects as health, education, improved nutrition, and so on are recognized as not only improving physical well-being and quality of life of rural people, but also indirectly enhancing the productivity of rural people and increasing their ability to contribute to the national economy) (World Bank 1975*b*).

Experience from most developing countries tends to suggest that rural development efforts have not been particularly successful in promoting rural welfare. It appears that progress has been hampered by, among other factors, very high rates of population growth—often over 2.5 per cent per annum—and associated demographic changes. Yet despite such experiences, very few of these countries have devised any policies that address themselves to the population changes. This chapter shows how population growth and associated demographic changes have substantially influenced the pattern of development in rural Malawi. Specifically, population growth and associated demographic changes in rural Malawi have affected smallholder agricultural development, mostly through changes in cultivated area, and this has altered income distribution among the rural populations.

The Trend of Population Growth in Malawi

Since the beginning of this century, the population of Malawi (which was estimated at 5.55 million in 1977) has been growing at an estimated annual

Table 8.1   Population growth trends in Malawi (1901–77 censuses)

Year	De facto	De jure	Intercensal average annual growth rate (per cent)
1901	n.a.	737,153	2.8
1911	n.a.	970,430	2.8
1921	n.a.	1,201,983	2.2
1926	1,263,291	1,293,291	1.5
1931	1,573,454	1,603,454	4.4
1945	2,049,914	2,183,200	2.2
1966	4,039,583	4,305,583	3.3
1977	5,547,460	n.a.	3.2
2000*	10,238,000*		2.7

Source: Malawi Government 1980*a*, vol. 3, table 6.1.
* Author's projection based on 2.7 per cent annual growth rate on the 1977 *de facto* population.

rate of 3.2 per cent. (Malawi Government 1980*a*). This implies that the Malawian population doubles almost every twenty-five years or so, and this seriously affects access to cultivable land and, consequently, income distribution.

Although the country has a total land area of 9.4 million ha, only 5.3 million ha (56 per cent of total) are rated as cultivable (Malawi Government 1980*b*). In addition, as Pike (1968) has observed, although Malawi's soils are fertile by African standards, their nutrient status shows wide differences. This implies that the cultivable land is unequally distributed regionally, and that there exists great diversity in farming patterns in the country.

This is partly because the customary land tenure system prevails in the Malawian smallholder agriculture subsector. This tenure system principally safeguards against landlessness, in that the land is communally held with the chief or some other traditional authority as the main custodian who allocates land to households and families for private use. Thus, as long as someone is a member of the society, he/she is entitled to a piece of land which can be passed on to heirs. The system is flexible in that the chief can reallocate land away from a household that has unused land to one that is in need of more land. As such, the amount of land a household cultivates tends to be dependent on the ability to cultivate the land, which in turn depends on the amount of labour (family and/or hired) which the household commands. Thus, in the face of rising population pressure on the land, it is mostly labour-deficient households that lose their unused land to those with more resources, and this implies that subsequent generations inherit very small pieces of land which may not provide enough for subsistence purposes in a situation of static technology.

This variation means that population growth and accompanying demographic changes have had differing impact on different parts of the country. Generally, it can be said that the rapid increase in population has resulted in growing pressure on the available cultivable land, and a number of indicators bear this out. For example, in a sample survey of agriculture undertaken in 1968/9, only 1.4 million ha of the cultivable area (26 per cent of cultivable land) were under permanent cultivation at the time (Malawi Government 1980*b*). However, according to the estimates of the land husbandry section of the Ministry of Agriculture and Natural Resources, by 1983 many areas in the country will have exhausted all the idle cultivable land. In addition, the 1968/9 survey indicated that the average holding size (on customary land) was 1.54 ha per household, with an average of 4.6 persons per household.

A more recent survey of smallholder agriculture conducted in 1980/1 has actually revealed that the average holding size has declined from the 1.54 ha recorded in 1968/9 to 1.17 ha, which represents a 24.6 per cent decline in a period of about 12 years (Malawi Government 1984). Again, according to the population census undertaken in 1966, the population density in the country averaged 43 persons per square kilometre, while the 1977 population census recorded an average population density figure of 59 persons per square kilometre, which represents a 37 per cent rise in 11 years (Malawi Government 1978). Clearly, pressure on the land is mounting, and without substantial improvements in technology to raise productivity on the land a substantial proportion of the smallholding community will abandon the land to search for alternative sources of income. This is already becoming evident in rural Malawi. This chapter investigates the impact of this rapid population growth and associated demographic changes on Malawi's rural development efforts in general, and how different socio-economic groups specifically are being affected, especially in terms of their access to land and other resources.

One drawback of census and household survey data in the Third World is that data are often highly aggregated. The most that one can do to analyse this data is to calculate averages based on the whole population.[1] This presupposes that the rural population is fairly homogeneous, and that analyses based on the average household are realistic. However, a closer investigation of the rural population reveals that there is quite a substantial degree of heterogeneity in its make-up. For example, there are substantial inter-household differences in such aspects as farming systems, resource availability, family sizes and compositions, and so on, all of which bear strongly on the households' income levels, and so on. In such a case, analyses based on the average household tend to mislead. The present writer contends that unless we disaggregate the rural population into a number of criterion-based typologies or categories, we cannot understand how different socio-eco-

---

[1] Very little is done in terms of breaking down the population into socio-economic typologies and basing the analysis on these typologies. Although the 1980/1 National Sample Survey data was broken down into Agricultural Development Divisions (ADDs), this is still a very high level of aggregation and cannot reveal any useful information on inter-household differences on socio-economic aspects.

nomic groups have fared as a result of our rural development efforts in the face of rising population pressure and associated demographic changes.

The main issue here is to examine access to land among the different socio-economic groups and what this has meant in terms of the income levels of the various groups. Drawing mainly from a number of studies conducted in rural Malawi, an attempt is made to indicate how access to cultivable land has affected the whole pattern of rural development in the country.

The Impact of Land Pressure on Malawi's Rural Development Efforts

In Malawi, as in most developing countries, rural development occupies a prominent position in the national development plans. Because over 90 per cent of the rural population in the country earn their livelihood from small-holder agriculture, the country's rural development efforts have tended to emphasize improving smallholder productivity, mostly by encouraging small farmers to use productivity-raising inputs and to follow improved crop and animal husbandry practices. As such, the extent to which each household or family can benefit from the rural development effort depends very much on the extent to which such a household is able to adopt such productivity-raising inputs and follow the recommended farming practices.

On the other hand, this is constrained by factors such as lack of resources to purchase the inputs, limited access to information, inadequate farm land, inadequate incentives associated with farm tenure arrangements and depressed farm prices, insufficient human capital, absence of equipment to relieve labour shortage (thus preventing timeliness of operations), chaotic supply of complementary inputs and inappropriate transportation infra-structure, and so on (World Bank 1981). The main strategy for Malawi's smallholder agricultural development has been to try to remove such con-straints by introducing facilities to provide farm credit, information, orderly supply of complementary and necessary inputs, and so forth. The main aim of this strategy is to enhance adoption of improved farming practices, which would lead to higher farm incomes.

A closer scrutiny of the strategy, however, indicates that it has not fully borne the expected results. Most noticeable is that there have been wide dif-ferences in adoption behaviour among the various socio-economic groups; that is, while some people have managed to improve their productivity and income positions, others have either stagnated or even deteriorated. This chapter proposes that the high population growth and its accompanying demographic changes have contributed to this phenomenon by affecting access to cultivable land.

In a study conducted by the present author in the Lilongwe Land Develop-ment Programme (LLDP) (Chipande 1983*b*), it was observed that three main factors interacted at the household level to determine the household's adoption behaviour.

Those people with large cultivated areas adopted most of the innovations introduced in the area (such as growing of fire-cured tobacco, hybrid and other types of improved maize, and ground nuts). Also, it was found that there was strong correlation between farm size and the amount of labour available to the household (both family and hired). Furthermore, there was a substantial relationship between farm size and access to credit (main source for inputs). The study also suggested that labour endowment of a household exercised an important influence on rural poverty and income distribution. For example, it was found that due to the Malawian land tenure system, labour-deficient households, such as female-headed households, tended to have small farms and limited access to credit, and could not therefore take up the innovations offered by the project.

Despite this situation, Malawi's rural development strategy continues to stress the importance of improving productivity among the country's small-scale farmers through enhancing adoption of innovation. Evidence tends to indicate, however, that productivity increases within the smallholder sector have not been substantial. For example, between 1964 and 1978 smallholder production (at constant prices) is said to have grown at about 3.9 per cent per annum. If measured against a population growth rate of about 3 per cent, a per capita growth rate in smallholder output was about 1 per cent per annum, whereas the annual rate of growth achieved in the estate sector during the same period was about 11.1 per cent (Ghai and Radwan 1980).

Importantly, the growth in agricultural output in both sectors was more a result of hectarage expansion than rising productivity. Thus, as population continued to grow, more marginal land was brought into cultivation—therefore actually resulting in declining land productivity. Therefore, the possibility for growth in smallholder output being higher than population growth was limited without a widespread adoption of innovations in the smallholder sector. Evidence from a number of studies in rural Malawi indicates that widespread adoption of innovations was never achieved (Chipande 1983*a*).

As a result, overall productivity in the smallholder sector remained low. This phenomenon, which largely results from the rising pressure on the land, has affected the pattern of the development of agriculture in two main ways:

1. An outflow of male labour from smallholder agriculture.
2. An increase in labour hiring in smallholder agricultural production.

What has been the impact of these two phenomena on Malawi's rural development strategy?

Labour Migration

As indicated, the rising pressure on the land has meant low productivity in smallholder agriculture. Because of the predominant position of smallholder agriculture in the rural areas, this has meant that economic opportunities in many rural areas are bleak. This has forced many able-bodied men to abandon small-scale agriculture in search of paid employment. The majority of these people end up on the estates (such as tobacco estates in Kasungu, Lilongwe, or Mzimba, or tea estates in Thyolo and Mulanje), whether as wage labourers or as tenants. Others have ended up in urban centres within the country; still others have followed the old tradition of migrating to South African mines and Zimbabwean factories and farms.[2] The implication of this migration is twofold. First, it has accelerated the growth of the estate sector *vis-à-vis* the smallholder sector. Second, it has meant that smallholder agricultural production has become predominantly a women's preoccupation, especially in those areas where male outmigration has been very pronounced.

A look at the population data in Malawi indicates that since the early 1970s, migration in Malawi (particularly male migration) has been largely rural–rural, rather than rural–urban. That is, many males have migrated from those rural areas with least economic potential (due to ecological factors, rising population pressure on the land, and similar factors), such as Phalombe, Chiradzulu, and other areas, to rural areas with highest economic opportunities, such as Lilongwe, Kasungu, Mchinji, and Mzimba. In the areas of high economic opportunity, expansion of tobacco estates has created opportunities either for wage employment or for the chance to become tenants on these estates (Chipande 1984). This migration becomes particularly evident when we compare the male–female ratios in the two types of areas before and after the tobacco expansion period, as Table 8.2 indicates.

As shown in Table 8.2, between 1966 and 1977 the areas of low agricultural potential such as Phalombe and Chiradzulu experienced a net outflow of males while the opposite is true for the tobacco expansion areas. This

Table 8.2   Comparison of male–female ratios among selected rural areas 1966 and 1977

	Males per 100 females		
	1966	1977	Change (per cent)
Low agricultural potential areas			
Phalombe*	91	88	− 3.3
Chiradzulu	89	86	− 3.3
Tobacco expansion areas			
Kasungu	86	108	+ 25.6
Mchinji	90	104	+ 16.6

Source: Malawi Population Census 1966 and 1977.
*For Phalombe area, the data used is for Nkhumba and Nazombe Traditional Authority areas.

2 International migration ceased to be of significant importance in the mid-1970s as a result of government policy.

is largely because in the early days of estate expansion the estate owners used to provide free transportation from places like Thyolo, Mulanje, Chiradzulu, and Zomba to recruit labour for the estates. In most cases, it was young men (in the 20–40 age group), often with little or no land, who took up this opportunity. Once on the estates, they settled either as direct labourers or as tenants. In most cases, they were given some plot of land on which to grow food for home consumption. Although many of these estates have now closed, due largely to bankruptcies, most of the young men have not returned to their places of origin. For example, as observed in a study conducted by the author in Mzimba district in 1985,[3] most of the young men who had migrated into the area to work on the estates were not offered free transportation back once the estates closed. As such, most of them sought alternative employment in the neighbourhood and/or ended up marrying and settling in the villages near the estates. Hence, outmigration has tended to be more permanent than temporary. This has created many female-headed households in the areas of outmigration, which has affected small-holder agricultural productivity.

When Malawi achieved political independence from Britain in 1964, it inherited an economic structure characterized largely by a European-owned, export-oriented estate sector and a predominantly low-productivity, subsistence-oriented smallholder sector. Thus, the post-independence agricultural development strategy was to increase productivity in the smallholder sector, which then accounted for over 85 per cent of total agricultural output and 53 per cent of total agricultural exports, and it provided over 90 per cent of the rural population with a livelihood.

Over the years, however, the smallholder sector has been overtaken due to external events. The estate sector doubled its share of total agricultural output and increased its share of monetary output from about 25 to 37 per cent between 1964 and 1978 (Ghai and Radwan 1980, 4). In addition, the small-holder sector's share of agricultural exports fell from 53.4 per cent in 1964 to 34.3 per cent in 1977 (Chipande 1983*a*, 74). This clearly shows that the estate sector has been growing at the expense of the smallholder sector. Also, the estate sector is still regarded as the main source of export earnings, which has resulted in favourable resource allocation for the sector (Kydd and Christiansen 1982).

The low productivity prevailing in the smallholder sector and the extractive policies followed by the parastatal (which enjoys a near-monopoly in marketing smallholder output[4]) have combined to cause very low returns to

[3] This observation was made when the author was conducting some research on income generating activities for rural women in Malawi on behalf of UNICEF/Ministry of Community Services, which was underway at the time of writing this chapter.

[4] Press Produce and a few licensed traders also handle some of the exportable smallholder output, whereas most of the food crops are disposed of through rural and urban markets. The impact of the smallholder crops pricing policy on rural development in Malawi has been covered in Chipande 1983*a*.

labour and other resources in the smallholder sector. This has resulted in an outflow of labour from smallholder production to estate production, as evidenced by the increase in wage employment in the agricultural sector between 1968 and 1980, from 44,100 to 82,700—a fourfold increase in a period of twelve years (Chipande 1983*a*). Most of this increase can be attributed to the fast expansion of tobacco estate production experienced in the 1970s. In Malawi, estate production is very labour-intensive. As such, its fast expansion was made possible only by the presence of a labour force prepared to shift from the relatively low-productivity smallholder sector to the relatively high-productivity estate sector. This shift of the labour force from one sector has had far-reaching and devastating implications for the development of the smallholder sector. Mainly, it has led to labour deficiencies in the smallholder sector, which have given rise to an increasing tendency to hire labour for agricultural production, thus creating a self-perpetuating spiral downward in smallholder productivity.

Labour Hiring

The strategy for improving productivity in the smallholder sector has been extremely ineffective. Small-scale farmers have been urged to adopt productivity-raising inputs and to follow modern husbandry methods. While extension services equipped the farmers with knowledge regarding how to handle these new inputs, obtaining credit was essential to implementing productivity-raising inputs. Yet evidence from studies conducted in rural Malawi (Chipande 1983*a*, Kydd 1982) suggests that input credit appears to have been provided to the farm population based on 'ability to repay'. That is, only those who convinced the project officials that they would be able to repay their credit at the end of the season stood any chance of obtaining the credit and thereby having a chance to use productivity-raising inputs to implement the extension advice.[5] Further, only those farmers who could produce a surplus beyond their subsistence needs (or had other sources of income to finance credit repayment) could qualify for credit. These farmers had large cultivated areas and households with large family labour forces (or they could afford to hire labour).

In the opposite situation, small and labour-deficient households (especially female-headed households) rarely adopted the input-raising innovations introduced by the project for two main reasons.

Their labour deficiency prevented them from having large cultivated areas, the main criterion for gaining access to credit.

Second, most of the innovations introduced by the project (such as modern crop husbandry methods that require proper and timely application

---

[5] Credit was mostly associated with the production of particular cash crops of tobacco, ground nuts, and improved maize varieties and not any other enterprises.

of fertilizers) require great labour intensity.[6] Therefore, those households unable to take up the innovations had low productivity, resulting in low income levels.

On the other hand, those households able to take up the innovations were able to raise their productivity and incomes substantially. These differences between the productivity and income of adopters and non-adopters eventually give rise to the emergence of a rural labour market in which the adopters hire the labour of non-adopters. How does this happen? Because the poor labour-deficient households do not use productivity-raising inputs, they get very little output from their farming operations. Unfortunately, such households usually have limited access to non-farm sources of income (given the low economic opportunities in the rural areas). This forces them to sell part of their output soon after harvesting, often at very low prices, to raise badly needed cash for such things as children's school fees, clothing, and so on. Most of these poor households grow subsistence crops rather than cash crops, and by selling part of their meagre output, they usually end up with food deficits. Therefore, many of these households buy food throughout the food-deficit period (December to January) from the better-off households, often at prices two to three times those they obtained at harvest time. Out of sheer necessity, these poor households must hire out their labour (the only resource they can sell at this time of the year) to better-off households, often at wages below the value (marginal revenue product) of their labour. In some cases, these labourers are paid in the form of food.

For the richer households, on the other hand, use of modern inputs greatly enhances their productivity and income. This induces them to take up even more innovations. Because these innovations are relatively more labour-intensive than the traditional crops, additional labour is required, usually even beyond the capabilities of their rather large family labour forces. Because these households operate on a higher production function than the poorer households, the productivity and income returns for labour on these farms exceed the returns on the labour in poorer farms. In this case, the innovating households gain from hiring the labour of non-innovating households at any wage rate below the marginal revenue of hired labour on their high-production-function farms. The non-innovating households, on their part, will only be able to raise their income levels by hiring out their labour to the innovators at any wage rate higher than the relatively lower marginal revenue of their labour on their own farms. Thus the differences in the production functions used by the two groups gives rise to a rural labour market, whereby the innovators hire the non-innovators at a wage rate that falls between the marginal revenue of labour on each type of farm (see Chipande 1983a, 220–30, for a more detailed discussion). The exact level at which the wage rate will settle will depend on the nature of the labour market

---

[6] Mellor (1966, 156), for example, makes this point that many technological innovations require added labour.

and on the bargaining power of each group. In the situation under review, because the hired labourers do it out of sheer necessity, they tend to be price-takers, and the wage rate tends to be low, which makes the whole situation rather exploitative.

The results of the survey undertaken in LLDP confirms the tendency toward increased labour hiring among the innovators. Here it was observed that 32 per cent of the sample reported hiring labour for agricultural work during the 1980/1 season. Of these, 82 per cent grew tobacco and improved maize varieties (the main innovations in the area). Or, put another way, about 50 per cent of those who adopted innovations hired non-family labour (the other 50 per cent having amply large family labour forces). In addition, it was observed that the poor non-adopting households relied largely on hiring out their labour to supplement their farm incomes (Chipande 1983*a*, 192–9).

Concluding Remarks

In the Malawi situation, rapid population growth has resulted in pressure on the land, which in turn has meant low productivity in the smallholder sector. This phenomenon has led to a number of interrelated aspects detrimental to smallholder agricultural development. First, it has led to an outflow of labour from agricultural smallholder production. This outflow of mostly male labour from the smallholder sector has resulted in labour deficiences in that sector. More important, it has meant that agricultural smallholder production is predominantly in female hands, and it has led to an increase in the proportion of labour-deficient (especially female-headed) households in the rural areas.

Unfortunately, evidence has indicated that both farm credit and extension services are biased against the labour-deficient households, since credit is given to those with large farms who mostly have large family labour forces and/or are able to hire labour, who are presumed to be able to grow a surplus over and above subsistence requirements. In addition, extension services and credit tend to go together, since extension personnel play a role in allocating credit and ensure that credit goes to those that accept advice. In turn, those that receive credit are closely supervised to ensure good results, thus enabling them to repay their credit loans.

The implication of this aspect is twofold. First, it means that unless the smallholder agricultural development strategy is changed in such a way that the bias against labour-deficient households in provision of productivity-raising inputs is reversed (say by offering smaller households smaller and lower-cost packages or less intensive innovations), the possibility for improving smallholder productivity in the face of increasing labour migration is negligible. Ultimately, the imbalance between the two subsectors of the agricultural sector inherited from colonial times will be made worse rather than corrected. The present policies will continue to allow the estate sector to receive a lion's share of the country's land and financial and human resources, in an attempt to maintain the viability of the country's total agricultural output. Thus the talk about using smallholder agricultural development as a rural development strategy becomes merely rhetoric.

Second, unless innovations become widely adopted among the smallholder community, the emerging pattern of innovation adoption (which depends on the better-off farmers adopting innovations largely with the help of the labour of the poorer non-adopting farmers) will accelerate the stratification of the rural societies. As already indicated, these poor households hire out their labour during the peak seasons, times when they badly need such labour on their own farms. This means that the following season such households are likely to end up with even lower outputs. Ultimately, their involvement in the rural labour market has a screw effect on them, driving them further into poverty after each successive round. The eventual result is that such households will be driven out of smallholder production to become full-time labourers on other people's farms and/or estates.

It must be pointed out that the shift of labour from small-scale agricultural production to estate production or the rising tendency for labour hiring with the smallholder sector, *per se*, may not necessarily lead to the impoverishment of the hired. In a situation where the wage rate is free to move in line with market forces, the movement of labour from a low-productivity subsector to a high-productivity subsector is likely to lead to a rise in the rural wage rate as the two subsectors compete for labour. This will in turn lead to an improvement in rural incomes. However, in the situation under review, this does not necessarily obtain, largely due to two factors. First, the government imposes a statutory minimum wage rate for unskilled labour in both rural and urban areas.[7] Although this rate has been reviewed periodically, the available evidence indicates that it has largely lagged behind the cost of living (Chipande 1983*a*, Ghai and Radwan 1980). At the same time, the statutory minimum wage policy does not seem to have been strongly enforced, with the result that in some cases some institutions (especially the newly established tobacco estates) do not strictly adhere to it. There have been instances whereby estate employees (especially female) have been paid below the minimum wage rate, largely because of the seasonal nature of their employment (Chipande 1985). Second, real farm-gate prices in smallholder agriculture have exhibited a downward trend since the late 1960s and early 1970s (Chipande 1983*a*). This invariably limits the extent to which hired labour in the subsector can be remunerated. Thus, faced with declining real wage rates and declining real farm-gate prices, those who end

---

[7] The main aim of such a wages policy is said to be to reduce rural–urban wage differentials among unskilled labourers and therefore reduce rural–urban migration. However, it also acts as an incentive for the estate sector to use labour-intensive methods of production.

up relying mostly on hiring out their labour to estates or other smallholders for a livelihood are likely to find themselves in persistent poverty.

With a fast-growing rural population and a continued expansion of the estate sector, the likelihood of such a phenomenon taking place in rural Malawi is great unless the current agricultural development strategy is changed to ensure that there is a proper balance between the estate sector and the smallholder sector, that innovation adoption becomes widely based among the smallholding community, and that the farm pricing and wages policies are structured in such a way that they ensure proper returns to labour.

References

Chipande, G. H. R. (1983*a*), 'Smallholder Agriculture as a Rural Development Strategy: The Case of Malawi', unpublished Ph.D. thesis, Department of Political Economy, University of Glasgow.

—— (1983*b*), 'Labour Availability and Smallholder Agricultural Development: The Case of the Lilongwe Land Development Programme (LLDP), Malawi', mimeo, WEP 10-6/WP61, ILO, Geneva.

—— (1984), 'Innovation Adoption Among Female-headed Households: The Case of Malawi', paper presented at a workshop on Conceptualizing the Household: Issues of Theory, Method and Application, held at Harvard University, 2–4 November 1984.

—— (1985), 'Women's Participation in the Thyolo–Mulanje Tea Estates', paper presented at the 8th Southern African Universities Social Science Conference, 15–17 July 1985, Zomba, Malawi.

Ghai, D. and S. Radwan (1980), 'Growth and Inequality: Rural Development in Malawi 1964–1978', World Employment Research Working Papers, Rural Employment Policy Research Programme, ILO, Geneva.

Kydd, J. G. (1982), 'Measuring Peasant Differentiation for Policy Purposes: A Report on a Cluster Analysis Classification of the Population of the Lilongwe Land Development Programme, Malawi, for 1970 and 1979', Department of Economics, Chancellor College, Zomba, Malawi.

—— and R. Christiansen (1982), 'Structural Change in Malawi since Independence: Consequences of a Development Strategy Based on Large Scale Agriculture', *World Development* 10(5).

Lipton, M. (1977), *Why the Poor Stay Poor: A Study of Urban Bias in World Development*, Temple Smith, London.

Malawi Government (1978), *1977 Preliminary Census Report*, National Statistical Office, Zomba.

—— (1980*a*), *Malawi Population Census, 1977, Final Report*, National Statistical Office, Zomba.

—— (1980*b*), *Malawi Statistical Yearbook, 1979*, National Statistical Office, Zomba.

—— (1984), *National Sample Survey of Agriculture 1980/81*, National Statistical Office, Zomba.

Massignon, N. (1973), 'Rural Development in the Third World: A Vital Need and a Challenge to Aid Programmes', OECD Observer 67.

Mellor, J. W. (1966), *The Economics of Agricultural Development*, Cornell University Press, Ithaca and New York.

Pike, J. G. (1968), *Malawi: A Political and Economic History*, Pall Mall Press, London.

World Bank (1975a), *Rural Development: Sector Policy Paper*, Washington.

—— (1975b), *The Assault on World Poverty*, Washington.

—— (1981), 'Adoption of Agricultural Innovations in Developing Countries: A Survey', World Bank Staff Working Paper no. 444, Washington.

9 Demographic Effects on Agricultural Proletarianization
The Evidence from India

PRANAB BARDHAN

Department of Economics, University of California, Berkeley

In the literature on rural development, agricultural proletarianization is usually associated with capitalist progress in agriculture. In the Marxist literature this is sometimes almost definitional: the capitalist mode of production being defined as one based on use of hired labour (with labour power bought and sold as a commodity), an increase in the proportionate importance of wage labour is taken as an index of capitalist progress.[1] Other Marxist writers associate capitalism not just with a particular form of extraction of surplus value, but also with the way the surplus is used in accumulation. In general it is presumed that even if proletarianization is not a sufficient condition for capitalism, capitalist progress dispossesses the small peasantry and drives them into wage labour. This chapter puts together some scattered evidence on the socio-economic and demographic factors that impinge on agricultural proletarianization and focuses on the relative importance of demographic factors in contrast to the others, noting their varying importance at different levels of disaggregation of data.

It is well known that India has had a significant agricultural wage labour force, going back a few centuries, even before the commercialization under colonial rule and the more recent demographic pressure that mounted in the early part of this century. Even today, intercountry comparisons in Asia suggest that factors other than demography are quite important. For example, the proportion of wage labourers in the agricultural work-force is much smaller than that of India in many countries of East Asia (like Korea,

---

[1] The Marxist philosopher Cohen actually asserts a logical connection between the development of a wage labour market and capitalist accumulation: 'if the producers are free labourers, production is for the sake of accumulating capital' (see Cohen 1978). In spelling out the steps in this argument, he writes: 'if labour is free and commodity production is well-established, there is competition between producing units. Competition between producing units imposes a policy of capital accumulation: a unit not disposed to increase the exchange-value at its disposal will lack the resources to prevail in competition.' I think the problem with Cohen's argument is the presumption of competitive markets. If there is territorial monopoly of landlords, and village markets are isolated and fragmented, a system of surplus extraction from local wage labour may survive for a long time without being forced by competition to serve the accumulation of capital.

Taiwan, and historically Japan and China) and of South-east Asia, including plantation agriculture (as in the Philippines and Thailand), even though in all of these countries cited, population density (specifically, the number of agricultural workers per hectare of arable land) is higher than in India.[2] No doubt the caste system in India, traditionally assigning the tasks of manual work in agriculture to some primarily landless low-caste persons and other ethnic groups, has historically played some role in differentiating the Indian case from other densely populated Asian economies. But it should also be noted that, in contrast to East and South-east Asia, the use of animal power in land preparation is critical in India.[3] Therefore, landless people without access to draught animals are often rationed out of the land-lease market to a much larger extent in India. This means that, compared to an East Asian counterpart, the Indian landless peasant is more often obliged to turn to wage labour for a living.

In order to isolate and quantify the effects of demographic factors from others, some statistical analysis at various levels of disaggregation are needed. First, in Indian agriculture, regional cross-section data strongly suggest that proletarianization does not necessarily go with capitalist development, and that it is more often associated with demographic pressure and inequality in land distribution. Let us cite some evidence from the data for 55 agro-climatic regions[4] covering most of rural India in the early seventies.

Table 9.1   Linear regression analysis of determinants of proportional importance of farm wage workers in male labour Force WMPROP across 55 regions in rural India, 1972–3

Explanatory variables*	Regression coefficient	Standard error	Significant at percentage level
1. Concentration ratio of land operated (CRLOP)	56.1526	13.6129	0.0
2. Gross capital expenditure on farm business per hectare of cropped area (GCEHA)	– 0.0404	0.0226	8.1
3. Density of population per km^2 (DENS)	0.0203	0.0103	5.4
4. Proportion of scheduled tribes in total male population (TRIBMP)	0.0866	0.0854	31.5
Constant term	– 16.3929	8.3551	5.5
$R^2 = 0.3667$; $F = 7.2$; no. of observations = 55			

* Dependent variable: percentage of farm wage labourers in rural labour force for males in the 15–59 age group (WMPROP)

[2] For summary evidence, see A. Vaidyanathan and A.V. Jose 1978, and A. Vaidyanathan, unpublished.

[3] The number of draught animals per hectare is much higher in South Asia than in South-east Asia. Vaidyanathan attributes this to the ecological fact that better seasonal distribution of moisture and lower temperature in East Asia make land preparation manageable with human labour, whereas in South Asia the hot dry summer hardens the soil to a degree that makes the use of animals necessary to break the ground and prepare the soil for sowing after the monsoon sets in.

[4] These regions have been classified by the National Sample Survey (NSS) by grouping districts that have similar population density and crop pattern. On average, a cluster of four or five homogeneous districts forms a region.

Table 9.2 Mean and standard deviation of variables used in Table 9.1

Variables	Mean	Standard deviation
1. WMPROP	21.48 per cent	10.32 per cent
2. CRLOP	0.6214	0.0981
3. GCEHA	38.34 (rupees)	54.90 (rupees)
4. DENS	178.04	124.14
5. TRIBMP	10.62 per cent	15.05 per cent

Sources: Same as those in Table 9.1
WMPROP is from the National Sample Survey 1973. CRLOP has been computed from the NSS *Land Holdings Survey* data for area operated by household operational holdings for 1970/1 (NSS 1971). GCEHA has been computed from the data on gross capital expenditure on wells and other irrigation sources, agricultural implements, and machinery as collected in Reserve Bank of India 1972, and the data on cropped acreage for 19 major crops for the average of the years 1970/1 to 1972/3 estimated by Bhalla and Alagh (1979). The data for DENS and TRIBMP are from the 1971 census.

From the NSS Employment and Unemployment Survey for 1972/3, we have estimates of the proportion of farm wage labourers[5] in the total rural work-force for males in the 15–59 age group (WMPROP). This proportion varies from as low as 0.45 per cent in the Outer Hills region of Jammu and Kashmir to about 39 per cent in Inland Eastern Maharashtra. In Table 9.1 are presented the results of a regression equation explaining regional variations in WMPROP, and in Table 9.2 are the mean and standard deviation for the dependent and independent variables.

The first thing to note in the regression results of Table 9.1 is that WMPROP has a significant *negative* association with gross capital expenditure on farm business per cropped hectare (GCEHA). In other words, proletarianization is in general *less* in regions of relatively high capital investment in agriculture. On the other hand, WMPROP is positively, and significantly, associated with the index of concentration of cultivated land (CRLOP) and the density of population (DENS). WMPROP is also positively, though weakly, associated with the proportion of tribal population (TRIBMP).

How does one interpret the results? Looking at the figures for WMPROP across the agro-climatic regions, one clear pattern is that in high-rainfall, naturally fertile areas population density is high, the proportion of landless or near-landless is large, wage labour is cheaper to hire, and the extent of proletarianization[6] is relatively high, even though these areas are not necessarily technologically progressive. Historically, these are areas of relatively secure wet agriculture where the mounting population is absorbed with an

[5] Farm wage labourers are defined as those whose principal source of income is from farm wage work by usual status.
[6] One may object to the term 'proletarianization' by pointing out that some of this wage labour may be unfree or 'bonded'. As we have discussed in Bardhan 1984, the incidence of bonded labour in Indian agriculture is quantitatively insignificant except in localized pockets. Cases of bonded labour are often associated with remote isolated areas. Our regional cross-section evidence suggests that in sparsely populated remote areas WMPROP tends to be low.

astonishing elasticity (a process described by Geertz as one of 'agricultural involution'), with continuous subdivision and multiplication of rights on land leading to elaborate hierarchies, and with the actual manual work of cultivation done only by those who are at the bottom of the social heap (such as ethnic groups like tribals). The large numbers of landless and the elaborate vertical layering of land rights also imply that in many of these areas inequality in land distribution is high, which, as the regression results indicate, is very strongly correlated with WMPROP. On the other hand, in sparsely populated areas with a lot of remote villages not well connected with centres of commercialization (which tend to commercialize the labour market as well), the extent of usage of wage labour is low. Capital investment is relatively large (and accelerated in recent decades) in the agriculture of Punjab, Haryana, and western Uttar Pradesh. But in these areas the land–person ratio is relatively high and the proportion of wage labour low. Unlike in the densely populated areas, the land hierarchy here is relatively simple and oriented to direct cultivation of land by owners (or owner-cum-tenants), and the farmers depend to a large extent on the labour of their extended families. Thus, the influence of demographic, ecological, and institutional features of a region on its degree of agricultural proletarianization may often be more important than the degree of capital accumulation *per se*.

Evidence from Household Data in West Bengal

The empirical results of the previous section were based on cross-section evidence from heterogeneous agro-climatic regions. In order to focus on the demographic factors, let us now turn our attention to a much more disaggregated set of data for a relatively homogeneous area where the ecological and institutional features are not as sharply different as in the interregional data. This data set relates to about 4,500 agricultural households in more than 500 villages in the state of West Bengal in 1977/8, as collected by the NSS. (For both a description and detailed statistical analysis of these household-level data, see Bardhan 1984.) This section briefly reports on some of the results relating to the effects of demographic factors on agricultural labour participation. In our analysis of the determinants of the variable, HOUM (that is, the number of person-days in the reference week for which men in the 15–60 age group in these agricultural households were hired out on farm work or reported seeking or being available for work), we find that the estimated partial[7] elasticity of HOUM with respect to an independent variable that we call LABFORM (the total number of such men in the household currently in the labour force) is 0.96 for the primarily agricultural wage labour households and 0.77 for the primarily cultivator households. This

---

[7] This elasticity is estimated, keeping the household land cultivated and other factors constant.

Table 9.3   LOGIT analysis of the probability of usual labour force participation by women in 15–60 age group in agricultural wage labour households in rural West Bengal, 1977–8

Explanatory variables	Estimated coefficient	Standard error
ADWOMEN* (number of adult women in household)	0.9801	0.0859
LABFORM* (number of adult men in the household who are currently in the labour force)	– 0.7367	0.0867
BAB* (number of children in 0–4 age group in the household)	– 0.2445	0.0689
SCHTRIBE* (dummy variable for households belonging to scheduled tribes)	1.3948	0.1653
SCHASTE* (dummy variable for households belonging to scheduled castes)	1.0013	0.1127
CHDOM* (number of children in household in 5–14 age group currently in domestic work)	0.2452	0.1154
VILPOP* (village size in thousands of population in 1971)	0.0023	0.0012
VWAGEM* (average daily wage rate in rupees for male agricultural labour in village in the reference week)	– 0.3764	0.0496
HUMNOWN* (dummy variable for unowned homestead)	0.8655	0.1561
AWAYM* (dummy variable for some male member of the household gone away for work)	0.3685	0.3094
CULTIVAT* (area cultivated by the household in acres)	– 0.2329	0.1082
EDM* (number of men with above-primary education level in the 15–60 age group in the household)	– 0.6384	0.2470

Likelihood ratio index = 0.2614; number of observations = 2,276

Notes: The data for the variables are from the household schedules of National Sample Survey 1973, except for VILPOP which are from 1971 census.
* Denotes a variable significant at 5 per cent level.

means that an increase in the number of adult male workers in the family increases the supply of labour to the farm wage labour market, but less than proportionately. We also find that the variable HOUM increases significantly, with an increase in the variable representing the number of dependents as proportion of the household size.

In the case of women in these households, their participation in the farm labour market in this area is rather low; therefore, variations in it may require some special analysis. In the primarily agricultural wage labour households in our sample for rural West Bengal in 1977/8, on average, only 3 out of 10 women in the 15–60 age group participated in the labour force by 'usual status' (defined as the status that prevailed over the major part of the preceding year). Because female labour participation is often a dichotomous decision (with women in the majority of these households deciding not to participate), we carried out a LOGIT analysis of the probability of such participation for these households. The results are reported in Table 9.3. For a detailed explanation of the estimated coefficients, see Bardhan 1984. This section focuses on the coefficients of only the demographic variables: ADWOMEN (number of adult women in the household), LABFORM (number of adult men currently in the labour force), BAB (number of children in the 0–4 age group in the household), CHDOM (number of children in the 5–14 age group currently in domestic work in the household), SCHTRIBE (dummy variable to indicate whether the household belongs to a scheduled tribe), SCHCASTE (dummy variable to indicate whether the household belongs to a scheduled caste), and VILPOP (the population size of the village in which the household is located).

From Table 9.3, it appears that the probability of labour force participation[8] by any woman in the agricultural wage labour households increases significantly with (*a*) an increase in the number of adult women in the household; (*b*) an increase in CHDOM (suggesting how children, usually female, doing domestic work ease the constraint on adult women in these households for work outside); (*c*), membership in scheduled caste and tribe categories (confirming the weakness of taboo on outside work for women in these socially disadvantaged groups); (*d*) a larger village population size (possibly suggesting that a larger village with its attendant commercialization may provide more opportunities for work and less-rigid taboos on female participation). This probability of female labour participation is lower, the larger the number of babies and very small children in the family (indicating the usual child-care constraint on female labour participation[9]), and the larger the number of adult men in the family who are in the current labour force (presumably indicating the usual income effect).

Intertemporal Evidence

In the preceding two sections, we have provided cross-sectional evidence at the regional and the household level of the importance of demographic effects on agricultural proletarianization and labour participation. Turning from cross-sectional evidence to fragmentary time-series data in rural India, the demographic effects on proletarianization seem quite strong. On the basis of Rural Labour Enquiry and NSS data, it seems that the proportion of rural households deriving a major part of their income from wage labour rose from around a quarter in 1964/5 to some 30 per cent in 1974/5 to about 37 per cent in 1977/8. Rural population during the 1960s and 1970s grew by about 45 per cent, while gross cropped area rose only by about 13 per cent. The data presented for the major states in Table 9.4 show that the proportion of agricultural labour households, while it grew quite fast in states like Punjab, Haryana, Gujarat, and Karnataka where the rate of growth of food-grain production was high, also rose substantially in states like West Bengal, Bihar, Tamil Nadu, Orissa, and Madhya Pradesh, where the annual

---

[8] Although this participation is in terms of all types of gainful work, the overwhelming part of it is in farm wage labour.

[9] I am, of course, assuming here that women's labour force participation behaviour itself does not affect their reproductive bahaviour.

Table 9.4   Rates of growth and changes in importance of agricultural labour households in major states

States	Annual percentage growth rate of rural population over 1961–81	Annual percentage growth rate in production of food-grain between the triennia ending in 1961/2 and in 1981/2	Agricultural labour households as percentage of total rural households		
			1964/5	1974/5	1977/8
Andhra Pradesh	1.6	2.1	31.4	35.8	41.4
Bihar	1.8	1.3	28.0	33.3	36.1
Gujarat	2.1	3.8	16.7	22.3	31.1
Karnataka	1.8	2.8	27.2	30.8	37.9
Kerala	1.8	1.1	28.2	27.4	27.0
Madhya Pradesh	2.0	0.6	20.3	21.8	27.9
Maharashtra	1.8	2.2	31.1	32.0	38.6
Orissa	1.7	1.3	24.7	30.1	37.1
Punjab–Haryana	1.6	5.5	14.4	15.8	21.6
Rajasthan	2.4	0.4	5.5	4.0	10.0
Tamil Nadu	1.4	1.3	28.0	38.1	39.3
Uttar Pradesh	1.8	2.1	13.9	15.8	18.1
West Bengal	2.1	2.0	25.4	30.9	35.8

Sources: The population growth rates are estimated from census data, the food-grain production growth rates from Ministry of Food and Agriculture data, the rural labour household data for 1964/5 and 1974/5 from *Rural Labour Enquiry* data (Government of India 1975), and those for 1977/8 from NSS 32nd Round data.

per capita rate of growth of food-grain production over the two decades was negative.

Over time, demographic pressure can, of course, have conflicting effects on the extent of proletarianization. On the one hand, by lowering the average size of per capita land, it drives more people to the wage labour market for livelihood or supplementary support; on the other hand, in a country where inheritance practices are characterized by land subdivision rather than primogeniture, some of the erstwhile labour-hiring large farmer households will reduce their labour hiring and may even pass from the rich farmer class to that of predominantly family farmer class. As the joint family system erodes over time,[10] this subdivision of land ownership gets translated more often into subdivision of cultivation, reducing the dependence on hired labour. On the other hand, in the demographic structure as the age distribution of the rural population shifts over time in favour of the young, this may tend to raise the proportion of agricultural labourers (the Indian census data suggest that the proportion of agricultural labourers among male rural workers is inversely related to age).

[10] Bhalla, on the basis of her study of rural Haryana, claims that the adoption of new technology, by raising income from cultivation per acre, has removed an economic constraint previously binding the joint family together (see S. Bhalla 1977).

In different areas, these demographic effects are either reinforced or counteracted by forces unleashed by the process of economic growth itself. There are several ways in which economic progress accelerates the process of agricultural proletarianization. Increased profitability of agriculture (along with protective land legislation) often leads to eviction of erstwhile tenants by landowners: given the terms and conditions of tenancy contracts for these insecure tenants, this, of course, in many cases only means a conversion of a semi-proletariat class into the full-scale proletariat. Then again some of the erstwhile small farmers and self-employed artisans are forced by new economic circumstances to join the ranks of agricultural labourers. In the case of farmers faced with a situation of credit rationing and paucity of capital, the new circumstances may mean a rise in the need for purchased inputs and credit intensity of cultivation, and in the case of artisans, the new circumstances may be a fall in demand for traditional crafts and disintegration of household industry, particularly as the demand pattern of the new rural rich shifts to mass-produced urban consumer goods and services.

Some scattered evidence supports this. The data on tenancy eviction are, of course, difficult to collect, but nevertheless findings (Bardhan and Rudra 1978) from a survey of 110 randomly selected villages in West Bengal in 1975/6 show that 81 per cent of the villages in the agriculturally more advanced areas reported an increase in tenant eviction, as opposed to 19 per cent in the backward areas. As for evidence on cultivation shifting away from small farmers, the all-India evidence is not clear. The NSS data (Sanyal 1977) do not show any sign of increased concentration of operational holdings at the national level between 1954 and 1971. But it is interesting to note from the same data that in the fastest-growing regions of Punjab and Haryana, the inequality in the distribution of cultivated land increased unambiguously (in the sense of Lorenz curve dominance) between 1960/1 and 1970/1. In 1970/1 in Punjab (including Haryana), even though only 9 per cent of rural households did not *own* any land, as many as 54 per cent did not *cultivate* any land; the corresponding percentages in 1960/1 were 12 and 39 respectively. Similar changes may also be noted in other high-growth states like Gujarat and Karnataka over the same period. Even for the country as a whole, two pieces of information may be relevant in this context. One is that on the basis of cross-sectional evidence (like that used in Tables 9.1 and 9.2) for 55 NSS regions in the early 1970s) the estimated correlation coefficient between the percentage of area irrigated and the proportion of rural households owning but not operating land is 0.52. This may suggest that in better-irrigated areas, the cost of cultivation may have driven some small land-owners away from farming. The second piece of evidence is from Rural Labour Enquiry data: the proportion of rural labour households *with land* as a proportion of all rural labour households in India increased from 43 per cent in 1964/5 to 49 per cent in 1974/5, possibly indicating an increased dependence of small landed households on wage labour. As for the declining importance of household industry, a comparison of the 1971 and 1981 census data suggests that the porportion of total male rural workers in household industry went down in almost all the states (except West Bengal and Tamil Nadu), subject to some problems of comparability in the definition of 'household industry' between the two census reports.

Finally, demographic pressure influences not merely the extent of proletarianization but also the composition of the agricultural wage labour force. If one distinguishes the casual wage labourers from those who are 'attached' to the employer in varying forms of long-term contract duration, one expects to see a positive association between demographic pressure and the importance of casual labour, the argument being that in areas where labour supply is relatively plentiful the employer need not bother to have long-term contracts with labour, because the employer is more assured of labour supply for peak agricultural operations.[11] In the aforementioned NSS interregional cross-section data, there is some evidence that there is a strong negative association between the importance of what NSS called 'regular' (as opposed to casual) farm labourers in 1972/3 in a region and the proportion of rural households that are landless or otherwise asset-poor.

References

Bardhan, P. (1984), *Land, Labour and Rural Poverty*, New York.
—— and A. Rudra (1978), 'Interlinkage of Land, Labour and Credit', *Economic and Political Weekly* 13 (February).
Bhalla, G. S., and Y. K. Alagh (1979), *Performance of Indian Agriculture: A Districtwide Study*, New Delhi.
Bhalla, S. (1977), 'Changes in Acreage and Tenure Structure of Land Holdings in Haryana, 1962–72', *Economic and Political Weekly* 12 (March).
Cohen, G. A. (1978), *Karl Marx's Theory of History: A Defense*, Princeton, N J.
The Government of India (1975), *Rural Labor Enquiry 1974–75: Final Report on Wages and Earnings of Rural Labor Households*.
National Sample Survey (1971), *Land Holdings Survey for 1970–71*.
—— (1973), *Employment and Unemployment Survey for 1972–73*.
Reserve Bank of India (1972), *All-India Debt and Investment Survey for 1971–72*.
Sanyal, S. (1977), 'Trends in Some Characteristics of Land Holdings: An Analysis for a Few States', *Sarvekshan, Journal of the National Sample Survey Organization* 1 (July/October).
Vaidyanathan, A., 'Impact of Development on Rural Wage Labour in India', unpublished.
—— and A. V. Jose (1978), 'Absorption of Human Labour in Agriculture: A Comparative Study of Some Asian Countries', in P. K. Bardhan *et al.* (eds.), *Labour Absorption in Indian Agriculture*, ILO/ARTEP, Bangkok.

[11] For a theoretical as well as a more detailed empirical analysis of this phenomenon, see Bardhan 1984, chapter 5.

Part IV

The Role of Frontier Expansion as a Safety-valve

10 Frontier Expansion, Agricultural Modernization, and Population Trends in Brazil

GEORGE MARTINE
International Labour Office, Brasilia

Much of the specialized literature has tended to view the relationship between population growth and distribution, agricultural development, and food supply within a rather narrow framework. Given a fixed supply of land, the problem has traditionally been one of discovering what land uses and which types of agricultural technology would permit the most adequate response to population pressure. *Ceteris paribus*, faster population growth would have to give rise to either greater intensity of land use and higher yields or to the appearance of Malthusian spectres. Within such a framework, the interaction among economic sectors receives little attention, as does the question of fit between agricultural development and the overall growth model.

At one time, such approaches may have been prompted by early images of underdeveloped countries: rigid, highly stratified social structures; absence of centralized politico-economic apparatus; high growth rates and density of population; widespread poverty; low crop yields; and no access to virgin lands. Be that as it may, a broader conceptual framework is required in order to understand the interplay between population and agriculture, at least in those countries which can still draw on relatively open land. Moreover, other factors alter the traditional equation in most of today's developing countries. The widespread adoption of the Green Revolution technological package has provoked various changes in social organization, depending on the manner of its introduction. The growth of centralized decision-making systems and the increasing interdependence among national, international, and multinational economic interests have all served to promote a tighter integration of agricultural activities with those of other sectors.

As a result, the indivisibility of agricultural problems from overall development strategies is increasingly emphasized. Mellor and Johnston's 1984 comment with respect to the food balance is illustrative in this respect:

The views expressed in this chapter are the sole responsibility of the author.

Our review leads to the conclusion that reduction of malnutrition and related manifestations of poverty requires a set of interacting forces, characterized as a ring, that link nutritional need, generation of effective demand for food on the part of the poor, increased employment, a strategy of development that structures demand towards goods and services which have a high employment content, production of wage goods and an emphasis on growth in agriculture.

The present chapter attempts to analyse the agricultural development of a land-rich country, Brazil, within the context of its overall growth strategies. Understanding the relationship between population dynamics, agrarian change, and access to land in this country requires an analysis of two main factors: frontier expansion and modernization of agricultural production. Endowed with enormous areas of open land, Brazil's frontier has traditionally shaped the structure of agricultural production while determining the size and composition of its rural population. Since the mid-1960s, however, the frontier's importance has been superseded by the sweeping, government-induced modernization of the agricultural sector. The purpose of this chapter is to discuss these two factors, their interrelations with demographic trends, and their implications for social development.

Frontier Expansion, Population Growth, and Agrarian Change

The Historical Significance of Frontier Expansion in Brazil

During the last half-century, frontier expansion has served two basic functions in Brazil. First, it has been systematically used as a safety-valve to release social tensions generated by stagnation, high population growth, and a rigidly stratified social structure in areas of traditional settlement. Second, frontier expansion in a land-rich country has permitted the increase of agricultural production without altering the land tenure system, the predominant forms of social organization, or the technological base of the prevailing structure.

As shown in Table 10.1, the land area under cultivation has increased greatly in Brazil between 1940 and 1980, as has the number of agricultural establishments and the size of the agricultural labour force. Broadly speaking, these changes have been produced by three major frontier movements.[1] These are differentiated from each other both in chronological and spatial terms: the settlement of the state of Paraná, of the centre-west region, and of the Amazon region. Paraná experienced heavy rural immigration from the mid-1930s to the mid-1960s. It benefited from several comparative advantages in terms of proximity to ports and to market centres, selectivity of settlers, quality of soil, and land tenure system. The central states of Goias,

[1] For elaborate discussions of these three frontier movements, see George Martine 1982*a*, 270–92; 1982*b*, 53–76; 1982*c*, 146–70.

Table 10.1   Selected indicators of agrarian change, Brazil, 1940–80

Date	Cultivated land area (thousand ha)	Number of agricultural establishments (in thousands)	Agricultural labour force (in thousands of persons)
1940	18,835	1,905	11,343
1950	19,095	2,065	10,997
1960	28,712	3,338	15,634
1970	33,984	4,924	17,582
1980	49,185	5,160	21,164

Source: Instituto Brasileiro de Geografia e Estatistica (IBGE), *Censos Agropecuários*, Rio de Janeiro (various years).

Mato Grosso do Sul, and Maranhão experienced accelerated growth during the 1950s and 1960s. Government investments in new cities, road-building, and other infrastructure stimulated the occupation of this region. Finally, more intensive settlement of the Amazon region was initiated in the early 1970s. For the first time, the federal government took the initiative of stimulating migration flows and attempting to organize the colonization of this huge area.

The importance of these three frontier movements for population redistribution, for the creation of economic opportunities, and for increasing agricultural production is undeniable. Taken together, in 1980 these three regions accounted for 30.6 per cent of all agricultural establishments (as compared to 16.2 in 1940), 28.5 per cent of all land area under cultivation (14.4 per cent in 1940), 31.1 per cent of the agricultural labour force (12.9 per cent in 1940), 28.8 per cent of the value of agricultural produce (13.4 per cent in 1940), and 21 per cent of the country's total population (12.5 per cent in 1940).

In short, the expansion of the frontier has clearly brought about the incorporation of new agricultural areas into the national economy, the diversification of its growth poles, and retarded urban concentration. Yet, although the occupation of new land areas is currently occurring at a faster pace than ever before, it is becoming an increasingly ineffective formula for the solution of social or production pressures. This relative loss of the frontier's significance is observable in (*a*) the declining demographic importance of rural areas on the frontier, (*b*) the slowing down in the rate of increase of its contribution to national production, and (*c*) the abridgement of the frontier's life-cycle in terms of intensive in-migration, stagnation, and out-migration.

Declining Demographic Importance of Rural Areas on the Frontier

Frontier growth is an increasingly urban phenomenon. That is, a decreasing proportion of all migrants to frontier areas end up on the land (see Table 10.2). The most likely explanation is that increasing limitations on access to

Table 10.2   Evolution of agricultural employment, according to the agricultural and demographic censuses, Brazil, 1960—80

| Year | Agricultural census | | | | Demographic census | |
| | Published data | | Revised data | | 'Economically active population' (in thousands) | Increase (per cent) |
	'Occupied population' (in thousands)	Increase (per cent)	'Occupied population' (in thousands)	Increase (per cent)		
1960	15.634	—	12.583	—	11.826	—
1970	17.583	1.2	12.402	– 0.1	13.090	1.0
1980	21.164	1.9	13.095	0.6	12.661	– 1.3

Note: See text for explanation of 'revised data'.
Source: IBGE, *Census Agropecuários e Demográficos* (various years).

land (as discussed in the next section) cause increasing numbers of would-be settlers and ex-settlers to seek survival on the urban fringe of frontier towns, where they wait hopefully to gain access to a plot of land. Whereas two-thirds of all population growth in Paraná at the height of its frontier movement (that is, during the 1940s and 1950s) occurred in rural areas, only 36 per cent of Amazonian growth during the 1970s accrued to rural areas. Indeed, seven medium-sized and large cities in that region (the state capitals plus Santarem) accounted for close to half of all Amazonian growth between 1970 and 1980.

In fact, however, the total number of rural migrants absorbed into the enormous Amazon region (comparable in size to Western Europe) was about half a million people, and fewer than 50 per cent of these were in official projects. This figure is roughly equivalent to the volume of migration absorbed by the Metropolitan Area (MA) of Belo Horizonte alone—and much smaller than the migration absorbed by either the MA of Rio de Janeiro or São Paulo.

Slowing Down in Growth of Rural Areas on the Frontier

Agricultural production as a whole increasingly centres in the older areas of settlement. New areas tend to contribute to production at a progressively slower rate. Several factors contribute to this diminishing importance of frontier expansion for agricultural production, despite the enormous increase in land area.

First, it is obvious that the availability of open, fertile lands close to important market centres has declined since the mid-1930s. Longer distances from market centres, lower soil fertility, and other more formidable natural obstacles to human occupation, *per se*, explain the delayed occupation of regions such as the Amazon. This does not mean that Amazonian lands cannot be profitably farmed, but simply that the relative mix of factors now favours more intensive exploitation of already occupied, centrally located lands, rather than extensive farming via incorporation of frontier areas. Rapidly rising transport costs tend to aggravate the distance factor and thereby promote a re-evaluation of the frontier's role in agricultural expansion.

Moreover, even the vast expanses of available unused land, particularly in the centre-west and Amazon regions (but also to a surprising extent in other regions), are the object of intense land speculation and disputes. Before the 1960s, much frontier expansion was led by small armies of *posseiros* or squatters, who invaded unoccupied regions and practised itinerant forms of slash-and-burn agriculture. For this class of small-scale, mostly subsistence farmers, land had a restricted utility value. After the land occupied by settlers had been cleared and basic infrastructure installed, it then acquired exchange value for a second wave of occupants—mostly cash crop farmers, ranchers, merchants, and other representatives of capitalist society—who took over the land by legal and/or forceful means. Thereupon, the *posseiros* move further into the interior, beginning a new cycle of appropriation–settlement–expropriation.

After 1965, however, the intense modernization of agricultural production (and the mechanisms utilized to promote it) caused agricultural land values to spiral upward. Speculation in land became extremely profitable and huge tracts of Amazonian land were also sold, bartered, or given outright to large national or multinational companies by the government— whether or not squatters and other small farmers were already on the land—in response to political pressures and under the justification that these organizations would promote a more 'rational economic exploitation' of the Amazon. Be that as it may, the upshot was the virtual elimination of the *posseiros* class and the closing down of the frontier to small-scale settlers.

Abridgement of the Frontier's Life-cycle

These factors (greater distance to markets, reduced soil fertility, soaring land values, restriction of access to land by small farmers, and the modernization of agricultural production) have converged to reduce continually the life-span of the typical frontier cycle in Brazil. This reduction can be seen clearly through an examination of the differential evolution of Brazil's three recent frontier experiences.

As observed earlier, the frontier region of Paraná benefited from an exceptional set of circumstances and, theoretically, should have produced a prosperous agricultural region, capable of absorbing and maintaining a large rural labour force. This did happen for the better part of thirty years, but Paraná's population growth rates, which had averaged close to 6.0 per cent yearly for the 1940–70 period, fell drastically to 0.94 per cent in the 1970s. More importantly, rural growth rates fell to – 3.5 per cent in the 1980s. This constitutes the sharpest reversion of demographic trends ever witnessed in Brazil; from a major locus of attraction, Paraná's rural areas suddenly became the nation's main supplier of out-migrants.

The occupation of the centre-west and Maranhão region, which peaked 10–15 years after Paraná, also shows clear signs of reversal. Although less dramatic than Paraná, out-migration from rural areas in the centre-west and Maranhão is now clearly the dominant trend. For instance, of 223 municipalities in Goias, 66 per cent had a growth rate smaller than that of the country and 35 per cent suffered an absolute decline in population between 1970 and 1980. Moreover, it is clear that the net rural loss from this region would have been much higher were it not for the attraction of the country's new administrative centre, Brasília. Although the characteristics of agricultural production in the states involved in this second stage of frontier development are markedly different from those of Paraná (particularly in terms of their more highly concentrated land tenure system), their experiences are not unrelated. First, the saturation of absorption possibilities is now clearly defined and traceable to changes in the structure of production. Second, the expulsion of rural residents is occurring mostly among small farmers, squatters, and sharecroppers.

Although the expansion cycle of the Amazon frontier is far from complete, it can be observed that its performance in terms of absorbing excess rural population has been rather disappointing, particularly in view of the region's size and the expectations surrounding its occupation. Indeed, net rural immigration to the region is relatively small. Though rural population figures in inaccessible states of the Amazon region are apt to suffer from severe underenumeration, even extreme corrections would not alter the fact that a large number of rural migrants who arrived less than ten years ago are already being forced to move on. First-hand evidence also testifies to the high turnover of settlers currently occurring in areas of very recent occupation such as Rondonia.

In short, the duration of the frontier cycle is being abruptly curtailed. In broad numerical terms, it could be stated that the entire cycle of intensive attraction–stagnation–expulsion lasted roughly 30–35 years in Paraná and 20–25 years in the centre-west and Maranhão regions, and will take only 10–15 years in the Amazon. Part of this truncation of the frontier cycle is due to the relative poverty of attractive features in more recent areas of attraction, but much is also due to the pervasive influence of the modernization process which, directly or indirectly, affects all forms of agricultural production in all areas of Brazil.

The Modernization of Agricultural Production in Brazil: Demographic and Social Implications

The Origins of Agricultural Modernization

Since the mid-1960s, much of Brazil's agricultural space has witnessed a switch from traditional and labour-intensive forms of production to more technologically advanced forms based on capital-intensive organization, improved varieties of seeds, and the massive adoption of the mechanical, biological, and chemical inputs which are part and parcel of the standard Green Revolution package. Although the specific implementation of the model has varied according to the historical conditioning factors and natural characteristics of each region, its influence has pervaded all major sectors of agricultural production.[2]

It is clear that the Green Revolution has provoked widely variable social transformation in different countries. In Brazil, the nature of social changes induced by it can ultimately be traced to its fit within a given model of economic development. The modernization of rural Brazil was conceived in the cities and implemented as a function of urban-industrial objectives, in consonance with a given political view of the society being designed. A first major determinant of this model was the decision (in the mid-1950s) to accelerate the process of industrialization via import substitution, by channelling major resources to the creation of a powerful domestic industrial sector.[3] This initiative eventually led not only to the loss of agriculture's traditional primacy, but also to the redefinition of its role within the economy; in addition to its responsibility in the production of foods, export crops, and raw materials for industry, agriculture was now looked upon as an important potential market for the budding industrial sector.

However, this reorientation of the economy, and thus of agricultural production, effectively took hold only after the instauration of a military regime in 1964. That government's ideology of conservative modernization via acceleration of industrialization, coupled with a more efficient centralized technocracy, deeply altered the traditional alliances and favoured modern or entrepreneurial forms of agricultural production. A sharp rise in international agricultural prices (registered in the 1968–73 period) provided additional stimulus at a crucial moment. Moreover, the development model which was adopted brought about a growing dependence on the foreign market; this in turn enhanced the need for increases in both production and productivity in the agricultural export sector.

The main instrument devised to ensure the viability of this new agricultural model was massive subsidized rural credit. Basic infrastructure was provided, and special programmes and other specific policies were implemented, but agricultural credit unquestionably constitutes the main thrust of government intervention in the agricultural sector after 1964. The number of credit contracts and the volume of resources doled out by government

---

[2] This section builds upon previous efforts by the present author, particularly Martine 1984, 69–98; see this publication for bibliography and additional data.

[3] Though the following discussion proceeds as if the major decisions concerning the choice of development style, and within it, of agricultural models in Brazil were autonomous and independent, this does not correspond to reality. Why, when, and how external influences affected the choice of development model, as well as the specific manner of implementing the agro-modernization, is itself a fascinating question—but one which would lead us far afield.

agencies soared. However, their social and spatial distribution acted as a catalyst in the disintegration of the existing structure.

Indeed, the bureaucratic operation of bank loans inherently favoured the concentration of resources into a small number of larger loans, given exclusively to farmers who could show title to the land and otherwise represented a low risk. Hence, rural credit became highly concentrated in the more developed central-south regions, in the hands of larger farmers, and for the production of capital-intensive, high-yield, export-oriented crops.

Consequences of Modernization

The impact of this modernization policy on the organization of agricultural production can be identified in terms of changes in the crop mix, in the technological base, in land values, and in the structure of land tenure. It also had a remarkable effect on employment, wages, labour relations, migration, and urbanization.

Crop Mix

In the post-1964 development scheme, because agriculture was to play an important role not only as a producer of raw materials but also as a potential

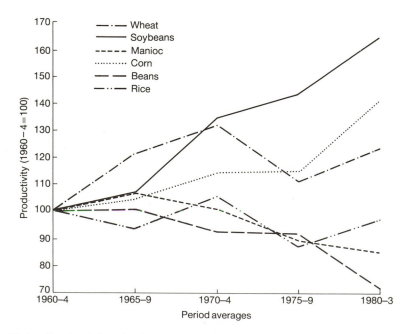

Fig. 10.1.   Productivity of selected crops, Brazil, 1960–80
Source: IBGE, Anuários Estatísticos (various years) and CEPABRO/MINAGRI/IBGE.

market for the industrialized sector, those crops regarded as 'dynamic' (that is, destined for export and/or agribusiness while also capable of generating a demand for agricultural machinery and inputs) were given preferential treatment in the provision of credit and other forms of incentives.

Not surprisingly, capital-intensive, export-oriented crops were favoured to the detriment of food staples such as rice, beans, and manioc. The area destined for food staples grew at a much slower rate than that given over to dynamic and/or export crops. Moreover, productivity of basic foodstuffs levelled off or declined, partly because of the lack of technological research and development, partly because these crops were relegated to less accessible or less fertile areas. Meanwhile, the productivity of soy beans, wheat, and corn (only 15 per cent of which is destined for human consumption within Brazil) increased significantly (see Figure 10.1).

Technology

The utilization of tractors serves as an indicator of the adoption of modern farm technology. Tractor usage increased significantly after the implantation of heavy industry in the 1950s. However, it was only in the 1970s—after the institutionalization of subsidized credit had begun to revolutionize farming methods—that the biggest boost in tractor usage was registered. Thus,

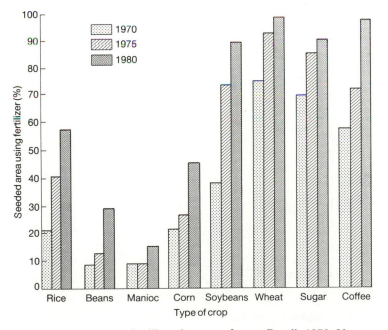

Fig. 10.2. Seeded area using fertilizer, by type of crop, Brazil, 1970–80

the number of tractors rose from 166,000 to 323,000 between 1970 and 75 and from 323,000 to 531,000 between 1975 and 1980.

By the same token, the utilization of chemical and biological inputs, although still low by developed-country standards, soared during the 1970s. The utilization of fertilizers serves as an indicator of this change. Available information, spanning the 1970–80 period, clearly shows that fertilizer use has multiplied with reference to practically all crops over the period; nevertheless, this increasing use is much greater in crops oriented to exports or agribusiness, such as soy beans, coffee, or wheat than in standard staples (see Figure 10.2). As could be expected, the use of fertilizers, preventives, and/or irrigation also varies directly with the size of agricultural establishments in all crops.

Land Tenure

Although a strict cause-and-effect relationship cannot be established, the innovations in credit structure and availability, as well as in the adoption of modern technology, evidently played an important role in accelerating the changes in the land tenure system observed during recent years. From the very early stages of its colonization process, Brazil had always been characterized by a tremendous concentration of land ownership. Sugar *engenhos*, cattle ranches, coffee *fazendas*, and so on, all functioned on large estates. As a result, land concentration in Brazil is, to this day, among the

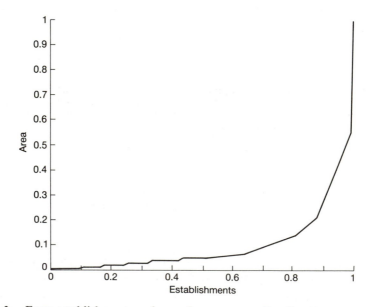

Fig. 10.3.   Farm establishments and area, Lorenz curve, Brazil, 1980
Source: IBGE (1980), *Censo Agropecuário*.

highest in the world, as can be seen from the Lorenz Curve in Figure 10.3. (see also Graziano de Silva and Hoffman 1980, 3–17).

Nevertheless, if one looks at the long-term trend, one notes a slow, clear-cut, progressive reduction in the size of farm establishments up to 1970. All categories of small farms up to 100 ha gained in the share of total farm land up to 1970, while all those of 1,000 ha or more showed a proportionate loss. Admittedly, these slow changes in no way added up to anything close to land reform, but at least the trend was in the right direction.

After 1970, however, the tendency towards a gradual reduction in average farm size was abruptly reversed; all other farm size categories lost relative importance due to the expansion of establishments in the category of 1,000 ha or more. Consequently, by 1980, small farms had a percentage share equal or below that which they had had in 1960 (see Figure 10.4).

The probable course of events that led to this concentration of land is not hard to reconstruct. Access to bank credit depends on the guarantee provided by land ownership. The more land owned, the easier it is to obtain credit and the greater the amount of obtainable subsidized loans. Land titling has always been extremely confused in Brazil and thus specialists in legal and/or forceful (even violent) land appropriation had a field day. Because credit was subsidized and loosely monitored, credit for land development

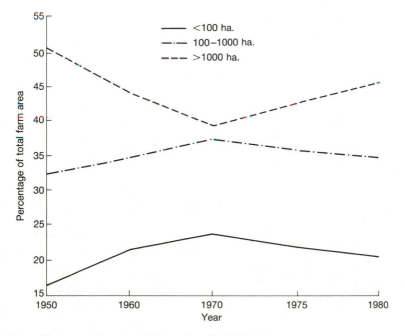

Fig. 10.4.   Farm area by size of farm, Brazil, 1950–80
Source: IBGE, *Censo Agropecuário* (various years).

was often used for the purchase of more land. In this process, smaller producers, who were less likely to possess a legal title, less familiar with banking and less attractive to bankers, less capable of fighting legal battles, and less capable of physically defending their properties or even of turning down apparently attractive cash offers to sell out their plots of land, were, predictably, the major victims. Land values, which had increased only by 130 per cent between 1966 and 1971, leaped by 2,000 per cent in the 1971–7 period, when rural credit was most readily available.

Employment and Wages

The same transformations in the structure of production that affected land ownership brought about profound changes in employment and wages. The main sources of data on these questions are the agrarian and demographic censuses. Published results of each diverge due to differing objectives and methodologies, but it is possible to reconcile them and spell out main tendencies.[4] Thus, uncorrected employment figures from the agrarian census show slow growth between 1960 and 1970 and somewhat faster growth between 1970 and 1980; meanwhile, the demographic census shows slow growth in the 1960s and an absolute reduction in the 1970s.

These discrepancies are, in large part, due to differences in concepts and definitions, particularly as concerns the secondary labour force (children under 15 and women). Thus, if we revise these data by simply substituting figures relating to this group in the agrarian census by the corresponding figures from the demographic census, then the discrepancies between the two sources are greatly reduced (cf. two middle columns of Table 10.2). The agrarian census still tends to show higher current employment figures, possibly because of double counting, the inclusion of temporary workers, and other factors related to the definition of employment.

In any case, the main point is that no matter which source is utilized, the best that can be said about the rate of growth of rural employment in the period under study is that it has been quite slow, particularly if compared to the population growth rate. Rural fertility during the 1960s and 1970s was roughly 50 per cent higher than that of urban areas, while mortality differences were relatively small. Thus, natural increase in rural areas tended to be considerably higher. As of 1985, natural growth rates at the national level were estimated at 2.1 per cent, while those of rural areas were around 2.6 per cent yearly. Overall, despite the massive incorporation of new land areas, employment growth rates of 1.0 per cent or less since the mid-1960s were equivalent to about one-third of the growth rate of the rural labour force.

However, perhaps the main consequence of the modernization process on employment has been its impact on the nature and stability of jobs in rural

---

[4] For a detailed analysis of these two sources, their discrepancies and convergences, see George Martine and Alfonso Rodríguez Arias 1985, 98.

areas: the tendency to transform rural labour into a wage-earning labour force. Traditionally, Brazilian agricultural techniques had always been intensive in human labour, as well as concentrated in large land-holdings. Much of the labour force for these large farms was provided either by share-croppers and other kinds of tenants who resided on the premises or by small farmers who regularly bartered their labour in exchange for goods and supplies. Thus, in 1950, such tenants composed 11.3 per cent of all individuals involved in agricultural activities. (This proportion more than triples if family-owned small farms are excluded from the total.)

It would be naive to attribute connotations of a harmonious, equitable, and prosperous social organization to pre-capitalist settings. Nevertheless, small farmers were, in the past, typically assured of a permanent subsistence base in terms of employment, food supply, and stability under the various traditional forms of self-provisioning agriculture and patronage. But soaring land values, modernization of production techniques, and the threat of labour legislation being applied to rural areas caused sharecropping to decrease rapidly. Between 1970 and 1980, this group suffered a reduction from 1,603,000 to 649,000. Thus, in 1980, only 2 per cent of the agricultural labour force were sharecroppers. By contrast, the number of wage workers, both permanent and temporary, has grown rapidly during the same period, according to all available sources of information. For instance, raw figures from the agrarian census showed an increase of 87 per cent in wage workers between 1970 and 1980. The recent ascendancy of temporary workers is particularly noteworthy. In the 1970s an increase of 1,242,000 such workers makes it, by far, the fastest-growing occupational group in agriculture.

Temporary work, of course, derives largely from the cyclical nature of agricultural activity and from the seasonal fluctuations in labour demands. In Brazil, the forms of temporary activity are highly variable among regions due to differences in climate, crop mix, land tenure systems, and modes of social organization. Nevertheless, it is clear that the most salient characteristics of temporary work are instability and unpredictability. In contrast to seasonal workers in some other countries, the *boia-frias* (agricultural workers recruited on a daily basis in towns) do not have access to a sequence of crops or harvests that would guarantee employment during a major part of the year and would thereby substitute the job stability of the former hybrid system of subsistence farming with occasional stints on the *patrão's* establishment.

In short, shifts in land tenure and in the structure of production have caused a profound change in the composition of the work-force and in labour relations, both by proletarianizing labour and by accentuating its instability. The impact of modernization on rural wages is also a pertinent concern in this context. It has been argued that the wages of unskilled labour have improved during the modernization process, although the data are somewhat contradictory, particularly as concerns the post-1979 period (see

Castro de Rezende 1985, Evenson 1982).

Whatever the validity of these data, it should be pointed out that arguments justifying the modernization of agriculture in terms of improvements in the wages of unskilled labour tend to disregard the changes in the nature of income which have followed the monetarization of the rural economy; given the pervasive reduction of 'in-kind' components, monetary wages would have to increase greatly in order for real income to remain constant. Such arguments also tend to neglect the inherently unstable nature of much rural employment, particularly among the fast-growing category of temporary workers; able-bodied male adults tend to avoid this type of work and seek other more permanent employment. But, most importantly, the figures neglect the fact that alleged improvements in rural wages affect only a proportion of the original rural population; the present contingent of rural workers is a selective residue after massive out-migration has taken place. This evolution of rural wages cannot be appraised without a simultaneous look at the changes in the labour force, in wages and in living conditions of both rural and urban populations. The real issue at stake is less how one restricted social stratum has fared than how society as a whole has been affected by rural transformations.

Migration and Urbanization

The lag between the pace of reproduction of the rural labour force and the growth of opportunities for work in agriculture lies at the root of the large recent rural exodus in Brazil; some 13.5 million rural migrants went to the towns and cities in the 1960s and 15.6 million more in the 1970s (Martine 1984, 87). Without delving deeply into this question, the data suggest that the tempo of rural migration since the mid-1960s is less a function of a region's relative level of poverty than of its stage with respect to the expansion of the agricultural frontier, on the one hand, and the penetration of modern capitalist forms of agricultural production, on the other.

Modernization of agriculture has intensified Brazil's urbanization process. True, the urban population had already grown approximately three times faster than the rural population throughout the 1940–70 period. However, the 1970s mark an important turning-point in Brazil's urbanization history when, for the first time, the country's rural population showed an absolute as well as a relative decline.

All city-size classes experienced accelerated growth, but data on residential distribution by size of place clearly reveal the progressive swelling of large cities to the detriment of rural areas, towns, and smaller cities. The greatest increase has been registered in cities of 500,000 or more, whose proportion of the total population jumped to 31.5 per cent in 1980. Moreover, during the 1970–80 decade Brazil's ten largest cities absorbed a number of people equivalent to 44 per cent of the country's total population growth in the period. Perhaps even more noteworthy is the fact that the three largest cities (São Paulo, Rio de Janeiro, and Belo Horizonte) absorbed 28 per cent of the national growth during the same period, with São Paulo alone accounting for 17 per cent of the total.

Rural–urban migration had been fairly substantial in Brazil since the 1930s, due to high rates of population growth within the context of a highly unequal land ownership system, rigid social stratification, and inefficient agricultural practices. But this migration has intensified since the 1970s, due to the penetration of industrial forms of production into agriculture. This shift in population toward large centres has aggravated a series of latent social problems in the cities.

The pressures which rapid urbanization exerts on urban services and infrastructure are well documented in the literature. Another area which will merit special attention, at least in the Brazilian context, is that of food availability. Indeed, the population transfers brought on by rural industrialization has had a double effect on the food problem in Brazil. Traditionally, and even to this day, basic staples are produced mostly by small farmers. The rural–urban displacement of many of these producers (whose own subsistence had also been previously assured on the farm) transformed massive contingents of food producers into mere consumers. In other words, the numerator in the producer–consumer ratio is reduced in the same proportion as the denominator increases. Huge increases in urban employment opportunities as well as significant improvements in income distribution would be necessary to offset this trend.

Summary and Discussion

Until recently, Brazil was widely reputed to have practically unlimited open lands for agricultural expansion at its disposal. Population pressures would not constitute a problem over the foreseeable future because progressive occupation of frontier regions would absorb excess manpower.

This simplistic view has been irrevocably shattered by events in the 1970s. Massive colonization programmes were conceived, but they were incapable of absorbing any notable proportion of the rural excess population. Indeed, the programmes themselves were soon scrapped due to lack of political support. Even more telling is the fact that previous successful experiences in frontier expansion suffered a complete reversal in migration trends, generating large contingents of out-migrants during the 1970s.

Both events are closely linked to the major changes that swept Brazil's agricultural system since the mid-1960s. The reorientation of Brazil's politico-economic model foresaw the transformation of the country's agricultural structure of production to increase its ability to consume industrial outputs and to produce certain 'dynamic' crops. Rural credit was the main (and effective) instrument used to carry out this project; modern farm technology was applied, and production was increased. In the process, land became even more highly concentrated, rural employment was reduced, rural–urban migration was intensified, and the availability of basic staples deteriorated.

In short, a given model of development, spawned largely within the context of urban-industrial aspirations, now reasserts itself against the cities in the form of rapid urban growth. Former subsistence farmers have been transformed into competitors for scarce urban employment and consumers of a reduced food supply.

This should not be construed as an argument in favour of a back-to-the-farm movement or in favour of birth control or other such panaceas. Increasing demands will inevitably be made on future agricultural production and productivity—for food supply, for energy-producing alternatives, and for reducing the balance-of-payments deficit—no matter which type of political regime were to assume control. The very demand for a progressive increase in agricultural production makes a return to pre-capitalistic models infeasible. Available technology favours a scale of production unattainable by small farmers working in isolation. The atomization of small farms greatly raises the costs of any subsidy or service to this group. In short, the international experience does not favour the hypothesis that production can be generically increased via small-scale producers.

On the other hand, the intensification of the current modernization process would produce an acceleration of rural–urban migration, of mass urban underemployment, and of food deficiencies. The absolute stock of rural population still amounts to 38 million people, with a natural growth rate of 2.6 per cent per annum. The agricultural frontier does not represent a viable alternative for the absorption of large numbers of labourers when in the context of land speculation and other restrictions on small farmers which have accompanied the modernization process. On the other hand, the current economic crisis has accentuated the precariousness of the larger Brazilian cities.

The challenge is thus to reduce the rate of expulsion of rural migrants while incorporating more effective production methods. Evidently, much could be done to improve productivity and living conditions among the large number of small subsistence farmers who still basically practise a form of hoe culture. The research into and adoption of capital-scarce, labour-intensive production methods, particularly in consumable crops, would be another steps in the right direction. Redistribution of land ownership, particularly of land which is being used ineffectively (or not at all) and a series of other measures commonly mentioned in such discussions would all help.

Nevertheless, it is essential to emphasize that the implementation of a few or all of such measures will not be of lasting significance unless they constitute part and parcel of a broad development strategy aimed at increasing employment and income redistribution in both rural and urban areas. Mellor and Johnston's comment regarding food supply, that 'the choice of development strategy is decisive in determining the level at which the food equation balances' (Mellor and Johnston 1984, 533), is equally applicable to the larger question of the balance between agricultural development and population growth. The subservience of agricultural changes and frontier expansion to a broader development model in Brazil since the 1970s reinforces the viewpoint that ultimately the decisions which determine the manner and course of sectoral production, as well as the allocation of the benefits of growth among different social strata, are essentially a political matter, not a technical one.

References

Castro de Rezende, G. (1985), 'Interação entre mercados de trabalho e razão entre salarios rural e urbanos no Brasil', Texto para Discussão no. 75, IPEA/INPES, March.

Evenson, R. E. (1982), 'Observations on Brazilian Agricultural Research and Productivity', *Revista de Economía Rural* 20 (3), July–September.

Graziano da Silva, J., and R. Hoffman (1980), 'A reconcentração fundiária', *Reforma Agraria* 10(6), 3–17.

Martine, G. (1982*a*), 'Recent Colonization Experiences in Brazil: Expectations versus Reality', in J. Balán (ed.), *Why People Move*, UNESCO, Paris.

—— (1982*b*), 'Expansao e retração de emprego na fronteira agricola', *Revista de Economia Política* 2(3), 53–76, São Paolo.

—— (1982*c*), 'Colonization in Rondonia: Continuities and Perspectives', in P. Peek and G. Standing (eds.), *State Policies and Migration*, Croom Helm, London.

—— (1984), 'Transformações recentes na agricultura e suas implicações sociais', in Agricultura: Rumos e Adjustamentos, Anais do XXII Congresso Brasiliero de Economia e Sociologia Rural, SOBER, vol. II.

—— and A. Rodríguez Arias (1985), 'A Evolução do emprego no campo', CNRH Working Papers, Brasilia.

Mellor, J. W., and B. F. Johnston (1984), 'The World Food Equation: Interrelations among Development, Employment and Food Consumption', *Journal of Economic Literature* 22 (June), 533.

Index